IET ENERGY ENGINEERING 193

Overhead Electric Power Lines

Other volumes in this series

Volume 1	**Power Circuit Breaker Theory and Design** C.H. Flurscheim (Editor)
Volume 4	**Industrial Microwave Heating** A.C. Metaxas and R.J. Meredith
Volume 7	**Insulators for High Voltages** J.S.T. Looms
Volume 8	**Variable Frequency AC Motor Drive Systems** D. Finney
Volume 10	**SF_6 Switchgear** H.M. Ryan and G.R. Jones
Volume 11	**Conduction and Induction Heating** E.J. Davies
Volume 13	**Statistical Techniques for High Voltage Engineering** W. Hauschild and W. Mosch
Volume 14	**Uninterruptible Power Supplies** J. Platts and J.D. St Aubyn (Editors)
Volume 15	**Digital Protection for Power Systems** A.T. Johns and S.K. Salman
Volume 16	**Electricity Economics and Planning** T.W. Berrie
Volume 18	**Vacuum Switchgear** A. Greenwood
Volume 19	**Electrical Safety: A guide to causes and prevention of hazards** J. Maxwell Adams
Volume 21	**Electricity Distribution Network Design, 2nd Edition** E. Lakervi and E.J. Holmes
Volume 22	**Artificial Intelligence Techniques in Power Systems** K. Warwick, A.O. Ekwue and R. Aggarwal (Editors)
Volume 24	**Power System Commissioning and Maintenance Practice** K. Harker
Volume 25	**Engineers' Handbook of Industrial Microwave Heating** R.J. Meredith
Volume 26	**Small Electric Motors** H. Moczala *et al.*
Volume 27	**AC–DC Power System Analysis** J. Arrillaga and B.C. Smith
Volume 29	**High Voltage Direct Current Transmission, 2nd Edition** J. Arrillaga
Volume 30	**Flexible AC Transmission Systems (FACTS)** Y.-H. Song (Editor)
Volume 31	**Embedded Generation** N. Jenkins *et al.*
Volume 32	**High Voltage Engineering and Testing, 2nd Edition** H.M. Ryan (Editor)
Volume 33	**Overvoltage Protection of Low-Voltage Systems, Revised Edition** P. Hasse
Volume 36	**Voltage Quality in Electrical Power Systems** J. Schlabbach *et al.*
Volume 37	**Electrical Steels for Rotating Machines** P. Beckley
Volume 38	**The Electric Car: Development and future of battery, hybrid and fuel-cell cars** M. Westbrook
Volume 39	**Power Systems Electromagnetic Transients Simulation** J. Arrillaga and N. Watson
Volume 40	**Advances in High Voltage Engineering** M. Haddad and D. Warne
Volume 41	**Electrical Operation of Electrostatic Precipitators** K. Parker
Volume 43	**Thermal Power Plant Simulation and Control** D. Flynn
Volume 44	**Economic Evaluation of Projects in the Electricity Supply Industry** H. Khatib
Volume 45	**Propulsion Systems for Hybrid Vehicles** J. Miller
Volume 46	**Distribution Switchgear** S. Stewart
Volume 47	**Protection of Electricity Distribution Networks, 2nd Edition** J. Gers and E. Holmes
Volume 48	**Wood Pole Overhead Lines** B. Wareing
Volume 49	**Electric Fuses, 3rd Edition** A. Wright and G. Newbery
Volume 50	**Wind Power Integration: Connection and system operational aspects** B. Fox *et al.*
Volume 51	**Short Circuit Currents** J. Schlabbach
Volume 52	**Nuclear Power** J. Wood
Volume 53	**Condition Assessment of High Voltage Insulation in Power System Equipment** R.E. James and Q. Su
Volume 55	**Local Energy: Distributed generation of heat and power** J. Wood
Volume 56	**Condition Monitoring of Rotating Electrical Machines** P. Tavner, L. Ran, J. Penman and H. Sedding
Volume 57	**The Control Techniques Drives and Controls Handbook, 2nd Edition** B. Drury
Volume 58	**Lightning Protection** V. Cooray (Editor)
Volume 59	**Ultracapacitor Applications** J.M. Miller

Volume 62 **Lightning Electromagnetics** V. Cooray
Volume 63 **Energy Storage for Power Systems, 2nd Edition** A. Ter-Gazarian
Volume 65 **Protection of Electricity Distribution Networks, 3rd Edition** J. Gers
Volume 66 **High Voltage Engineering Testing, 3rd Edition** H. Ryan (Editor)
Volume 67 **Multicore Simulation of Power System Transients** F.M. Uriate
Volume 68 **Distribution System Analysis and Automation** J. Gers
Volume 69 **The Lightening Flash, 2nd Edition** V. Cooray (Editor)
Volume 70 **Economic Evaluation of Projects in the Electricity Supply Industry, 3rd Edition** H. Khatib
Volume 72 **Control Circuits in Power Electronics: Practical issues in design and implementation** M. Castilla (Editor)
Volume 73 **Wide Area Monitoring, Protection and Control Systems: The enabler for smarter grids** A. Vaccaro and A. Zobaa (Editors)
Volume 74 **Power Electronic Converters and Systems: Frontiers and applications** A.M. Trzynadlowski (Editor)
Volume 75 **Power Distribution Automation** B. Das (Editor)
Volume 76 **Power System Stability: Modelling, analysis and control** A.A. Sallam and B. OM P. Malik
Volume 78 **Numerical Analysis of Power System Transients and Dynamics** A. Ametani (Editor)
Volume 79 **Vehicle-to-Grid: Linking electric vehicles to the smart grid** J. Lu and J. Hossain (Editors)
Volume 81 **Cyber-Physical-Social Systems and Constructs in Electric Power Engineering** S. Suryanarayanan, R. Roche and T.M. Hansen (Editors)
Volume 82 **Periodic Control of Power Electronic Converters** F. Blaabjerg, K. Zhou, D. Wang and Y. Yang
Volume 86 **Advances in Power System Modelling, Control and Stability Analysis** F. Milano (Editor)
Volume 87 **Cogeneration: Technologies, optimisation and implementation** C.A. Frangopoulos (Editor)
Volume 88 **Smarter Energy: From smart metering to the smart grid** H. Sun, N. Hatziargyriou, H.V. Poor, L. Carpanini and M.A. Sánchez Fornié (Editors)
Volume 89 **Hydrogen Production, Separation and Purification for Energy** A. Basile, F. Dalena, J. Tong and T.N. Veziroğlu (Editors)
Volume 90 **Clean Energy Microgrids** S. Obara and J. Morel (Editors)
Volume 91 **Fuzzy Logic Control in Energy Systems with Design Applications in MATLAB®/Simulink®** İ.H. Altaş
Volume 92 **Power Quality in Future Electrical Power Systems** A.F. Zobaa and S.H.E.A. Aleem (Editors)
Volume 93 **Cogeneration and District Energy Systems: Modelling, analysis and optimization** M.A. Rosen and S. Koohi-Fayegh
Volume 94 **Introduction to the Smart Grid: Concepts, technologies and evolution** S.K. Salman
Volume 95 **Communication, Control and Security Challenges for the Smart Grid** S.M. Muyeen and S. Rahman (Editors)
Volume 96 **Industrial Power Systems with Distributed and Embedded Generation** R. Belu
Volume 97 **Synchronized Phasor Measurements for Smart Grids** M.J.B. Reddy and D.K. Mohanta (Editors)
Volume 98 **Large Scale Grid Integration of Renewable Energy Sources** A. Moreno-Munoz (Editor)
Volume 100 **Modeling and Dynamic Behaviour of Hydropower Plants** N. Kishor and J. Fraile-Ardanuy (Editors)
Volume 101 **Methane and Hydrogen for Energy Storage** R. Carriveau and D.S.-K. Ting
Volume 104 **Power Transformer Condition Monitoring and Diagnosis** A. Abu-Siada (Editor)
Volume 106 **Surface Passivation of Industrial Crystalline Silicon Solar Cells** J. John (Editor)

Volume 107 **Bifacial Photovoltaics: Technology, applications and economics** J. Libal and R. Kopecek (Editors)
Volume 108 **Fault Diagnosis of Induction Motors** J. Faiz, V. Ghorbanian and G. Joksimović
Volume 110 **High Voltage Power Network Construction** K. Harker
Volume 111 **Energy Storage at Different Voltage Levels: Technology, integration, and market aspects** A.F. Zobaa, P.F. Ribeiro, S.H.A. Aleem and S.N. Afifi (Editors)
Volume 112 **Wireless Power Transfer: Theory, technology and application** N. Shinohara
Volume 114 **Lightning-Induced Effects in Electrical and Telecommunication Systems** Y. Baba and V.A. Rakov
Volume 115 **DC Distribution Systems and Microgrids** T. Dragičević, F. Blaabjerg and P. Wheeler
Volume 116 **Modelling and Simulation of HVDC Transmission** M. Han (Editor)
Volume 117 **Structural Control and Fault Detection of Wind Turbine Systems** H.R. Karimi
Volume 119 **Thermal Power Plant Control and Instrumentation: The control of boilers and HRSGs, 2nd Edition** D. Lindsley, J. Grist and D. Parker
Volume 120 **Fault Diagnosis for Robust Inverter Power Drives** A. Ginart (Editor)
Volume 121 **Monitoring and Control Using Synchrophasors in Power Systems with Renewables** I. Kamwa and C. Lu (Editors)
Volume 123 **Power Systems Electromagnetic Transients Simulation, 2nd Edition** N. Watson and J. Arrillaga
Volume 124 **Power Market Transformation** B. Murray
Volume 125 **Wind Energy Modeling and Simulation, Volume 1: Atmosphere and plant** P. Veers (Editor)
Volume 126 **Diagnosis and Fault Tolerance of Electrical Machines, Power Electronics and Drives** A.J.M. Cardoso
Volume 128 **Characterization of Wide Bandgap Power Semiconductor Devices** F. Wang, Z. Zhang and E.A. Jones
Volume 129 **Renewable Energy from the Oceans: From wave, tidal and gradient systems to offshore wind and solar** D. Coiro and T. Sant (Editors)
Volume 130 **Wind and Solar Based Energy Systems for Communities** R. Carriveau and D.S.-K. Ting (Editors)
Volume 131 **Metaheuristic Optimization in Power Engineering** J. Radosavljević
Volume 132 **Power Line Communication Systems for Smart Grids** I.R.S. Casella and A. Anpalagan
Volume 139 **Variability, Scalability and Stability of Microgrids** S.M. Muyeen, S.M. Islam and F. Blaabjerg (Editors)
Volume 145 **Condition Monitoring of Rotating Electrical Machines** P. Tavner, L. Ran and C. Crabtree
Volume 146 **Energy Storage for Power Systems, 3rd Edition** A.G. Ter-Gazarian
Volume 147 **Distribution Systems Analysis and Automation, 2nd Edition** J. Gers
Volume 152 **Power Electronic Devices: Applications, failure mechanisms and reliability** F. Iannuzzo (Editor)
Volume 153 **Signal Processing for Fault Detection and Diagnosis in Electric Machines and Systems** M. Benbouzid (Editor)
Volume 155 **Energy Generation and Efficiency Technologies for Green Residential Buildings** D. Ting and R. Carriveau (Editors)
Volume 157 **Electrical Steels, 2 Volumes** A. Moses, K. Jenkins, P. Anderson and H. Stanbury
Volume 158 **Advanced Dielectric Materials for Electrostatic Capacitors** Q Li (Editor)
Volume 159 **Transforming the Grid Towards Fully Renewable Energy** O. Probst, S. Castellanos and R. Palacios (Editors)
Volume 160 **Microgrids for Rural Areas: Research and case studies** R.K. Chauhan, K. Chauhan and S.N. Singh (Editors)
Volume 166 **Advanced Characterization of Thin Film Solar Cells** N. Haegel and M. Al-Jassim (Editors)
Volume 167 **Power Grids with Renewable Energy Storage, Integration and Digitalization** A.A. Sallam and B. OM P. Malik
Volume 172 **Lighting Interaction with Power Systems, 2 Volumes** A. Piantini (Editor)
Volume 905 **Power System Protection, 4 Volumes** Electricity Training Association (Editor)

Overhead Electric Power Lines

Theory and practice

Surajit Chattopadhyay and Arabinda Das

The Institution of Engineering and Technology

Published by The Institution of Engineering and Technology, London, United Kingdom

The Institution of Engineering and Technology is registered as a Charity in England & Wales (no. 211014) and Scotland (no. SC038698).

First published 2021

The Institution of Engineering and Technology
Michael Faraday House
Six Hills Way, Stevenage
Herts, SG1 2AY, United Kingdom

www.theiet.org

British Library Cataloguing in Publication Data
A catalogue record for this product is available from the British Library

ISBN 978-1-83953-311-2 (Hardback)
ISBN 978-1-83953-312-9 (PDF)

Typeset in India by MPS Limited

Expressing gratitude and sincere respect to
Parents, Teachers and all Well-wishers,
this work is dedicated
To
The Lotus Feet of Shri Shri Narayana

Contents

About the authors

Surajit Chattopadhyay (Ph.D., CEng, FIE(I), MIET) is an associate professor in the Department of Electrical Engineering in Ghani Khan Choudhury Institute of Engineering and Technology, India. His interests include power line installation, electric power quality, fault diagnosis, protection and signal analysis. He has authored/co-authored 3 books and more than 130 research articles in international and national journals and conferences, edited 1 book and received many awards and best papers' recognition. He is a member of the IET Communities Committee, South Asia, and former hon. secretary (2013–16) and YP-chair (2012–13), IET Kolkata Network.

Arabinda Das (Ph.D., CEng, FIE(I), FIETE) is a professor at the Department of Electrical Engineering, Jadavpur University, India. His research interests are power transmission, distribution, fault diagnosis, protection and power quality. He has published more than 85 research articles in the field of electrical machines and power systems in national and international journals and conferences and received the Railway Board Prize, the Union Ministry of Energy – Department of Power Medal, the Corps of Electrical and Mechanical Engineers Medal from The Institution of Engineers (India), among other awards.

Preface

Overhead electric lines share major part of power system network. In the last few decades, a lot of technological progress has been observed in power transmission as well as in the fields of material science, foundation technology and construction methods. These have brought changes in overhead line planning, construction and implantation. Moreover, environmental and geophysical aspects have become more important in the modern world. Keeping this in mind, a need to discuss all important aspects of overhead electric power lines considering the theoretical background and need in practice has become necessary at present scenario.

This book, Overhead Electric Power Lines: Theory and practice, has been written keeping all the relevant considerations of overhead lines in mind. After providing a brief introduction, fundamentals of power system relevant to the overhead lines have been presented followed by construction of different types of line supports, foundation and mechanical sag of line conductor with tension. Corona occurring in high-voltage line has been presented from design point of view. Then, the other two major components of overhead lines, overhead line insulators and conductors, have been discussed in detail. Different types of earthing along with protection against lightning and other surges have been presented. Insulation coordination has been discussed. At the end, different considerations for planning installation, commissioning and management have been discussed thereafter.

All the chapters have been presented in lucid manner covering theoretical and practical aspects.

Attempt has been made to correlate all theoretical aspects from design, installation and operational points of view to meet the need of advanced students, researchers and professionals working in this area. Authors would like to receive constructive criticism and suggestions from the readers and professionals for future modifications.

Surajit Chattopadhyay
Arabinda Das
Kolkata, India

Acknowledgements

In presenting the book 'Overhead Electric Power Lines: Theory and practice', we have received supports from different people and societies and sincerely acknowledge all the support obtained for the book.

We express our sincere thanks to Dr. Christoph von Friedeburg, Sr. Commissioning Editor for his continuous engagement in giving shape to the book in its present form. We also thank other supporting staffs of IET book publications.

We express our gratitude to our respected teacher Professor Samarjit Sengupta, past Chairman of IET Kolkata Network for his encouragement in the learning process.

We express our sincere thanks to Mr. Shekhar Sanyal, Country Head, IET India, his staff members and Mr. Kapil Khanna, past Chair and Mr. R N Rajpoot, Chair and all members of IET Communities Committee – South Asia for providing good platform to carry out and present the work.

We are thankful to all the members of executive committee of IET Kolkata Network.

We thank Dr. Tamal Roy, Dr. Debopoma Kar Ray and Dr. Aveek Chattopadhyaya of IET Kolkata Network and Dr. Santanu Chattopadhyay of NSOU, for their support at various stages.

We acknowledge the support received from Mr. Suvojit Basu for photography and express our sincere thanks to him.

Last but not the least; we are grateful to our parents and thankful to our children, spouses and all other family members for holding the patience during writing the book.

Surajit Chattopadhyay
Arabinda Das
Kolkata, India

"This book focuses on Electric Overhead Lines and professionally written by two established electrical experts (authors: Dr. Surajit Chattopadhyay and Dr. Arabinda Das) having vast experience in the Electrical Power Systems.

The most important part of this book is emphasizing advanced electrical technology and relevant engineering techniques on overhead lines (currently rarely available in the market).

This book is definitely for electrical power systems advanced students, professionals and researchers."

Lim Yew Kee CEng, MIET
Chairman
IET Internet of Things Technical Network
The Institution of Engineering and Technology (IET)
United Kingdom

"It is with great pleasure, I see new book of Dr. Surajit Chattopadhyay and Dr. Arabinda Das on 'Overhead Electric Power Lines: Theory and practice'. While there are quite a number of books on power systems, books focused on overhead power lines are few. I hope this book will prove to be a valuable treasure trove of knowledge and application in the area for students, researchers and practitioners in equal measures."

Shekhar Sanyal
Country Head and Director IET Services (India)

Chapter 1
Introduction

This chapter provides brief introduction of the book. After mentioning focus area and basic aspects of overhead electric power lines, it presents chapters at a glance.

1.1 Focus

With the advancement of technology, worldwide power demand is increasing. Power consumption has become the measure of technological progress of a country. With this, power system network has undergone various changes in terms of different technical parameters. Diversity has entered in generation, transmission and distribution. Transmission and distribution are still based on overhead lines mostly in AC form and partly in DC form. Power engineers are to cope with this ever changing complexity of the overhead lines. Changes in the overhead line technology are rapid and gradual aiming for long-term sustainability. This book attempts to present all aspects of overhead electric power lines to fulfil the need of advanced students and professionals working in this field. It focuses on fundamentals, line support, corona, line insulators, conductors, earthing, lightning and surge protection, insulation coordination, planning, commissioning and management.

1.2 Overhead lines

Overhead electric lines share major part of the transmission and distribution in power system. Technology involved in overhead electric power lines involves electrical as well as civil and mechanical engineering. In addition to that, management and performance audit take their role in smooth and effective implementation of the overhead line project. To deal with overhead lines, prior knowledge of power system fundamentals is necessary. Then, careful survey considering all relevant factors is needed for route selection. Design is done considering regional aspect, weather and geophysical feature along with electro-mechanical consideration. Physical work starts with foundation work by civil engineers. Then supports are installed followed by erection. Selections of tower type, conductors and insulations are very important. Safety and protective measured are required at each stage of the work. Insulation coordination is important for safety and reliable operation. Overhead line project involves planning, commissioning followed by operation and maintenance.

Wood, concrete and steels are in use for construction of support. However, comparatively use of wood is decreasing, whereas use of lattice tower has increased to a great extent in last few decades. With the advancement of material technology, design aspect of line insulators and selection conductors have undergone changes. Use of silicone composite insulators is increasing. On the other hand, different conductors are coming with greater current carrying capacity at better mechanical specification. These changes are introducing an option of uprating of line conductors and also replacement of old overhead lines by new lines.

1.3 Voltage level

Primary consideration for overhead line starts with nominal or operating voltage level at which the line is expected to operate. Voltage level for transmission is made as high as possible considering other factors to reduce loss. Therefore, transmission of electric power is found at high voltage or extra high voltage AC or high voltage DC. However, distribution of power is made medium or low voltage levels. Voltage level of overhead lines guides selection of tower dimension, conductor specification and insulation coordination. Tower dimension is again correlated with the type of foundation.

1.4 Safety measure

As overhead lines are exposed to atmosphere, they face various electro-physical abnormalities. Some problems may come from climatic variation. Therefore, protective measure, precaution and safety guidelines must be followed to achieve sustainable performance from the overhead lines. Special care is taken to provide protection against lightning, travelling waves and other over voltages.

1.5 Chapters at a glance

Considering various aspects of overhead lines, the book starts main discussion with fundamental of power system in Chapter 2. This chapter describes fundamentals of power systems. Different line parameters are presented. Resistance, inductance and capacitance have been derived. Skin and proximity effects have been explained. Classification of power line has been done followed by modelling of different parts of power system. Two-port network models of transmission lines have been presented, wherefrom transmission parameters, characteristic impedance, image impedance, surge impedance loading, etc. have been derived. Wave propagation properties have been shown. Comparisons of AC overhead lines with DC lines and underground lines have been provided. Modelling of synchronous generator and transformer has been described and then voltage regulation has been discussed with respect to transmission system. After that, different useful voltage compensation methods have been described.

In Chapter 3, different aspects of line support, foundation and mechanical sag of transmission line have been presented. It describes the design aspects with respect to distribution system and then with respect to transmission system. Different types of line supports have been presented. Design features of wooden pole, steel pole, concrete pole and lattice tower have been discussed. Different types of foundation methods have been presented for different line support. Both mono-block and compact foundations have been presented. Tests for supports have been highlighted. Mechanical sag with tension for overhead lines has been derived, and effects of ice and wind on sag have been described.

Chapter 4 deals with corona associated with high voltage or extra high voltage overhead lines. Its relationship with electric field is described. Critical corona is described. Power loss occurred due to corona in overhead transmission lines has been presented. Frequency and voltage dependencies of corona loss have been explained followed by consideration of other factors.

Line insulator plays an important role in overhead lines. Chapter 5 describes different aspects of overhead insulators. Properties of line insulators and materials used for manufacturing overhead line insulators have been presented. Different types of insulators used in overhead line application have been discussed. Requirement of string has been mentioned and then distribution of voltage across different discs of string has been shown followed by determination of string efficiency. Design features, fittings and selection factors of insulators for overhead application have been presented. Failure and testing of overhead line insulator have been presented. Different useful standards for this aspect have been referred.

Tower load is decided by line conductor along with effects of wind and ice. Power transmission capacity is also related with conductor. Chapter 6 deals with conductor used on overhead electric lines. It describes electrical, thermo-mechanical and other properties of line conductors. The conductor classification has been done followed by their description. Stranded conductors, bundled conductors are described. Materials used in different types of conductors have been presented. Different design features of hollow conductor, conductor with optical fibre cable, phase conductor, earth conductor have been presented. Jumpers' category has been mentioned. Covered conductors or overhead cables have also been discussed. Common stringing methods followed by different electrical properties have been presented. Ground clearance, stringing chart, conductor vibration, damper, spacer, etc. have also been described.

Earthing is important for safety measures. Chapter 7 deals with earthing and earth wire used in overhead electric lines. Effect of electric current on human body has been mentioned. Resistance properties of soil have been described. Measurement techniques for earth resistance have been presented. Different earthing procedures and materials suitable for earthing have been presented. Limitation of ungrounded system and advantages of earthed system have been presented. Fault current for different earthing systems have been described. Harmonic suppression system and earthing transformer have been described. Common grounding practices have been mentioned. Different properties and design features of earth wires and their selection guidelines have been presented. Some common practices for tower earthing have been provided.

Lightning and other over voltages are common in overhead lines and the line needs to be protected from them. Chapter 8 deals with lightning protection practices followed in overhead lines. It describes lightning strokes, its formation and characteristics. Return lightning discharge or return stroke and multiple strokes have been discussed. Then different protection schemes against lightning have been discussed. Earth wire or sky wire has been described. Different types of lightning arresters have been described. Wave propagation in transmission line has been explained with the help of characteristic impedance, propagation constant, etc. Different overvoltage assessment techniques or tests have been described.

For safe and reliable operation of overhead lines, coordination of insulation is must for overhead lines. As overhead lines operate at different voltage levels and are connected with other parts of power system network, in Chapter 9, knowledge of power frequency nominal voltage levels has been presented. Then lightning of other transient overvoltages have been presented. Concept of basic insulation levels has been provided mentioning its necessity. Factors of insulation coordination have been mentioned. Coordination with respect to insulation level of apparatus has been shown. Coordination with arresters has been described. Insulation coordination between overhead lines and connected substations has been presented followed by some useful standards for farther study.

In overhead line project, route selection, commissioning, operation and maintenance are different parts of job. Chapter 10 starts with route selection followed by general considerations of understanding of purpose, resource and object, covered area, geographical and geological diversity, political map, climatic statistics, cost study, etc. Different aspects of establishing new lines, alternate lines, line conversion have been discussed. Linking with underground lines has been discussed. Different aspects of planning and construction like survey, planning, design, foundation, installation, erection, commissioning, responsibility and issues, check-up, test, energization, etc., supervision, quality assurance in commissioning and commencement of operation have been discussed. Operation and maintenance, post-commissioning planning and management, operation management, maintenance management, asset management, risk management have been discussed. Uprating, upgrading and extension of lines have also been discussed thoroughly.

Chapter 2
Transmission line fundamentals

This chapter describes fundamentals of electric power lines. Different line parameters are presented. Resistance, inductance and capacitance have been derived. Skin effect and proximity effect have been described. Based on line parameters, classification of power line has been done followed by modelling of different parts of power system. A two-port network model of transmission lines has been presented. Transmission parameters, characteristics impedance, image impedance, surge impedance loading, etc. have been presented. Useful parameters of wave propagation have been shown. Comparison of AC overhead lines with DC lines and underground lines has been made. Modelling of synchronous generator and transformer has also been presented. Voltage regulation has been discussed with respect to transmission system. Different useful voltage compensation methods have been described.

2.1 Introduction

A typical power system consists of generation, transmission and distribution. Purpose of transmission line is to transmit generated electric power from generation side to distribution area. Electric power transmission line may be built up by overhead bare conductors or by underground cable. Most of the transmission lines are overhead line, whereas underground transmission lines are installed mainly in the areas where population density is high. Transmission lines are distributed over wide area and are connected with buses of wide voltage ranges. Voltage range depends on the area, type of transmission lines, supply and loads. In reality, all parameters of transmission lines are distributed in nature; however for simplicity of modelling of transmission lines of short and medium distances, parameters are assumed as lumped.

2.2 Classification of lines

Electric power transmission line is classified into following three categories:

1. Short transmission line (STL)
2. Medium transmission line (MTL)
3. Long transmission line (LTL)

In STL, line resistance and inductance are considered. But capacitance of STL is very small and neglected in analysis. Normally, line length of STL is less than 50 km. In modelling of STL, parameters are represented by lumped parameters.

MTL has considerable capacitance along with resistance and inductance. However, amount of the line capacitance is very small with respect to inductance. In modelling all parameters are considered as lumped, and the line is represented by either T or Pi model. Line length of MTL is normally in between 50 km and 100/150 km.

LTL has very line capacitance. Sometimes capacitive effect becomes higher than inductive effect. Resistance is much smaller than inductance and capacitance, and hence, resistive effect is often neglected in analysis. Modelling and analysis of LTL is done considering distributed nature of capacitance and inductance.

2.3 Line parameters

Electric power lines are characterized by resistance, inductance and capacitance. All these three parameters are distributed in nature. However, they are considered lumped for modelling of lines of short and medium length. In LTLs, parameters are considered distributed as they are in reality. Line parameters are discussed in following subsections.

2.4 Resistance

Resistance of a material where current is uniformly distributed can be determined from the following formula:

$$R = \rho \frac{l}{A} \tag{2.1}$$

where l is the length of the conducting material, A is the cross-sectional area of the conducting material, ρ is the resistivity of the conducting material.

2.5 Inductance

Inductance of a transmission line conductor forms due to the flux linkage which occurs both internally and externally. Thus, the total inductance of a conductor can be mathematically expressed as

$$L_T = L_{in} + L_{ex} \tag{2.2}$$

where, L_{in} is the inductance occurred due to internal flux linkage and L_{ex} is the inductance occurred due to external flux linkage.

2.5.1 Inductance of a conductor due to internal flux linkage

Let us consider a conductor of radius r and assume that the conductor is carrying current I which is uniformly distributed in the conductor. In this conductor, consider circular cross section of radius x as shown in Figure 2.1 and current I_x flows through it. I_x can be determined using Ampere's circuital law as follows:

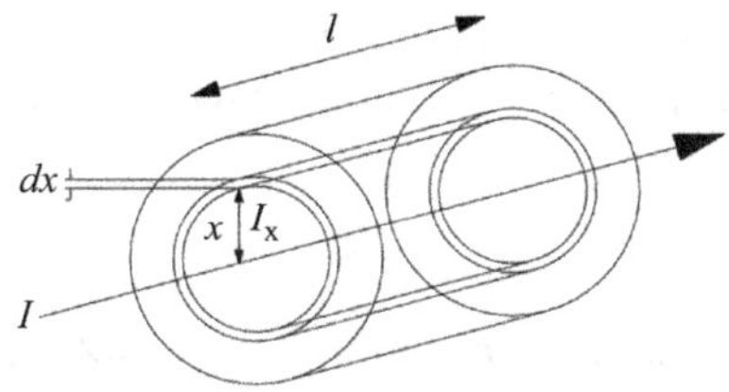

Figure 2.1 Cross-sectional view of a conductor

$$\oint H \cdot dl = I_{en} \tag{2.3}$$

where H is the magnetic field intensity corresponding to circle of radius x=H_x (say), I_{en} is the current enclosed by the circle of radius x=I_x.

Therefore,

$$\oint H_x \cdot dl = I_x \tag{2.4}$$

Now as the current is uniformly distributed, I_x can be expressed in terms of I as follows:

$$\begin{aligned} I_x &= \frac{\pi x^2}{\pi r^2} I \\ &= \frac{x^2}{r^2} I \end{aligned} \tag{2.5}$$

Therefore, from (2.4) and (2.5),

$$\begin{aligned} \oint H_x \cdot dl &= I_x \\ &= \frac{x^2}{r^2} I \end{aligned}$$

or,

$$\begin{aligned} 2\pi x H_x &= \frac{x^2}{r^2} I \\ H_x &= \frac{x}{2\pi r^2} I \end{aligned} \tag{2.6}$$

Therefore, magnetic flux density corresponding to magnetic flux intensity is given as follows:

$$\begin{aligned} B_x &= \mu H_x \\ &= \frac{\mu x}{2\pi r^2} I \end{aligned}$$

where μ is the permeability of the conductor.

Let us consider an elementary ring at radius x of width dx and also consider an elementary area dA formed by elementary distance dx in the radial direction and length 1 m as shown in Figure 2.1, i.e.

$$dA = 1 \cdot dx = dx$$

Therefore, flux distribution produced in the area dA due to flux density B_x can be written as follows:

$$\begin{aligned} d\phi &= \text{flux density} \times \text{area} \\ &= B_x \times dA \\ &= \frac{\mu x}{2\pi r^2} I dx \end{aligned} \tag{2.7}$$

Therefore, flux linkage created by $d\phi$ in the circle of radius x is written as follows:

$$\begin{aligned} d\varphi &= \frac{\pi x^2}{\pi r^2} d\phi \\ &= \frac{x^2}{r^2} d\phi \\ &= \frac{x^2}{r^2} \times \frac{\mu x}{2\pi r^2} I dx \\ &= \frac{\mu x^3}{2\pi r^4} I dx \end{aligned} \tag{2.8}$$

Therefore, total flux linkage can be written as follows:

$$\begin{aligned} \varphi_{in} &= \int d\varphi \\ &= \int_0^r \frac{\mu x^3}{2\pi r^4} I dx \\ &= \frac{\mu I}{8\pi} \end{aligned} \tag{2.9}$$

Therefore, line inductance developed by internal flux linkage φ_{in} can be written as follows:

$$\begin{aligned} L_{in} &= \frac{\varphi_{in}}{I} \\ &= \frac{\mu}{8\pi} \end{aligned} \tag{2.10}$$

The previous equation shows that inductance created by internal flux linkage is constant and independent of radius of the conductor. It depends only on the permeability of the conductor which can be written as follows:

$$\mu = \mu_r \mu_0 = 1\mu_0 = 4\pi \times 10^{-7}\text{H} \tag{2.11}$$

(Considering, $\mu_r = 1$.)

Therefore,

$$L_{in} = \frac{\mu}{8\pi} = \frac{1}{2} \times 10^{-7}\ \text{H/m} \tag{2.12}$$

2.5.2 Inductance of a conductor due to external flux linkage

Let us consider a conductor of radius r and assume that the conductor carries current I which is uniformly distributed in the conductor. Outside of this conductor, consider circular cross section of radius x as shown in Figure 2.2 and current I_x is flowing through it. I_x can be determined using Ampere's circuital law as follows:

$$\oint H \cdot dl = I_{en} \tag{2.13}$$

where H is the magnetic field intensity corresponding to circle of radius x=H_x (say), I_{en} is the current enclosed by the circle of radius x=I_x=I.

Therefore,

$$\oint H_x \cdot dl = I_x = I \tag{2.14}$$

or,

$$2\pi x H_x = I \qquad H_x = \frac{1}{2\pi x} I \tag{2.15}$$

Therefore, magnetic flux density corresponding to magnetic flux intensity is given as follows:

$$B_x = \mu H_x = \frac{\mu}{2\pi x} I \tag{2.16}$$

where μ is the permeability of the conductor.

Let us consider an elementary ring at radius x of width dx and also let us consider an elementary area dA formed by elementary distance dx in radial direction and length 1 m as shown in Figure 2.2, i.e.

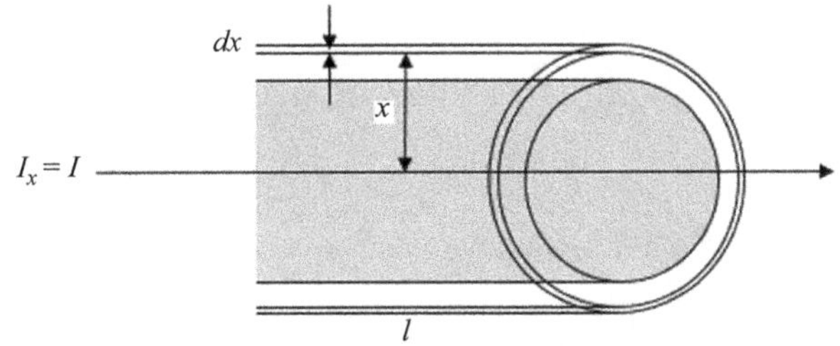

Figure 2.2 Circle of radius x enclosing conductor carrying current I

$$\begin{aligned} dA &= 1 \cdot dx \\ &= dx \end{aligned} \tag{2.17}$$

Therefore, flux distribution produced in the area dA due to flux density B_x can be written as follows:

$$\begin{aligned} d\phi &= \text{flux density} \times \text{area} \\ &= B_x \times dA \\ &= \frac{\mu}{2\pi x} I dx \end{aligned} \tag{2.18}$$

Therefore, flux linkage created by $d\phi$ in the circle of radius x is written as follows:

$$\begin{aligned} d\varphi &= d\phi \\ &= \frac{\mu}{2\pi x} I dx \end{aligned} \tag{2.19}$$

Therefore, the total flux linkage can be written as follows:

$$\begin{aligned} \varphi_{ex} &= \int d\varphi \\ &= \int_r^D \frac{\mu}{2\pi x} I dx \\ &= \frac{\mu I}{2\pi} \ln \frac{D}{r} \end{aligned} \tag{2.20}$$

Therefore, line inductance developed by external flux linkage φ_{ex} can be written as follows:

$$\begin{aligned} L_{ex} &= \frac{\varphi_{ex}}{I} \\ &= \frac{\mu}{2\pi} \ln \frac{D}{r} \end{aligned} \tag{2.21}$$

The previous equation shows that inductance created by external flux linkage depends on the radius of the conductor:

$$\mu = \mu_r \mu_0 = 1\mu_0 = 4\pi \times 10^{-7} \mathrm{H} \tag{2.22}$$

(Considering, relative permeability, $\mu_r = 1$.)
Therefore,

$$\begin{aligned} L_{ex} &= \frac{\varphi_{ex}}{I} \\ &= \frac{4\pi \times 10^{-7}}{2\pi} \ln \frac{D}{r} \\ &= 2 \times 10^{-7} \ln \frac{D}{r} \ \mathrm{H/m} \end{aligned} \tag{2.23}$$

2.5.3 *Total inductance of a conductor*

Total inductance of a conductor can mathematically be written as follows:

$$L_T = L_{in} + L_{ex} \tag{2.24}$$

where L_{in} is the inductance due to internal flux linkage, L_{ex} is the inductance due to external flux linkage.

Now,

$$\begin{aligned} L_{in} &= \frac{\mu}{8\pi} \\ &= \frac{1}{2} \times 10^{-7}\ \text{H/m} \end{aligned} \tag{2.25}$$

$$\begin{aligned} L_{ex} &= \frac{\varphi_{ex}}{I} \\ &= \frac{4\pi \times 10^{-7}}{2\pi} \ln \frac{D}{r} \\ &= 2 \times 10^{-7} \ln \frac{D}{r}\ \text{H/m} \end{aligned} \tag{2.26}$$

Therefore,

$$\begin{aligned} L_T &= \frac{1}{2} \times 10^{-7} + 2 \times 10^{-7} \ln \frac{D}{r} \\ &= 2 \times 10^{-7} \left(\frac{1}{4} + \ln \frac{D}{r} \right) \\ &= 2 \times 10^{-7} \ln \frac{D}{r\, e^{-1/4}} \\ &= 2 \times 10^{-7} \ln \frac{D}{r'} \end{aligned} \tag{2.27}$$

where,

$$r' = re^{-1/4} \tag{2.28}$$

Thus, the total inductance of a conductor of radius r is equal to the inductance due to external flux linkage of a conductor of radius $r' (= re^{-1/4})$.

2.5.4 *Inductance of three-phase transmission line having three conductors placed symmetrically*

Consider a three-phase transmission system made of symmetrically placed three conductors as shown in Figure 2.3. Let the distance between any two conductors be D, and the radius of each conductor be R.

Let the current in R, Y and B phases be I_R, I_Y and I_B, respectively. Flux linkage in phase R will be created by I_R (self) and by I_Y and by I_B (mutual).

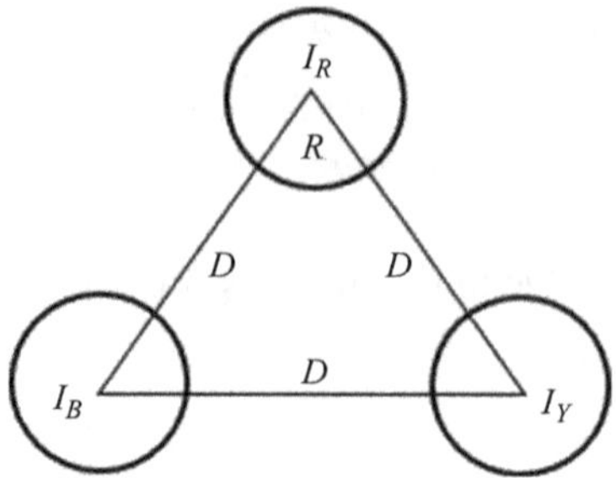

Figure 2.3 Three symmetrically placed conductors

Therefore, flux linkage in phase R can be written as follows:

$$\begin{aligned}\varphi_R &= 2\times 10^{-7}\left(I_R \ln\frac{1}{R} + I_Y \ln\frac{1}{D} + I_B \ln\frac{1}{D}\right)\\ &= 2\times 10^{-7}\left\{I_R \ln\frac{1}{R} + (I_Y + I_B)\ln\frac{1}{D}\right\}\end{aligned} \tag{2.29}$$

If the system is balanced,

$$I_R + I_Y + I_B = 0 \tag{2.30}$$

or,

$$I_Y + I_B = -I_R \tag{2.31}$$

Therefore,

$$\begin{aligned}\varphi_R &= 2\times 10^{-7}\left\{I_R \ln\frac{1}{R} - I_R \ln\frac{1}{D}\right\}\\ &= 2\times 10^{-7}\left\{I_R \ln\frac{1}{R} + I_R \ln D\right\}\\ &= 2\times 10^{-7} I_R \ln\frac{D}{R}\end{aligned} \tag{2.32}$$

Therefore, inductance in R phase can be written as

$$L_R = 2\times 10^{-7} \ln\frac{D}{R} \tag{2.33}$$

2.5.5 Limitation of three-phase three conductors placed unsymmetrically

Let us consider that a balanced three-phase supply is fed to a transmission line which is made of three conductors placed unsymmetrically as shown in Figure 2.4. Also assume that the line is connected to a three-phase load such that the same current is flowing through each phase, i.e.

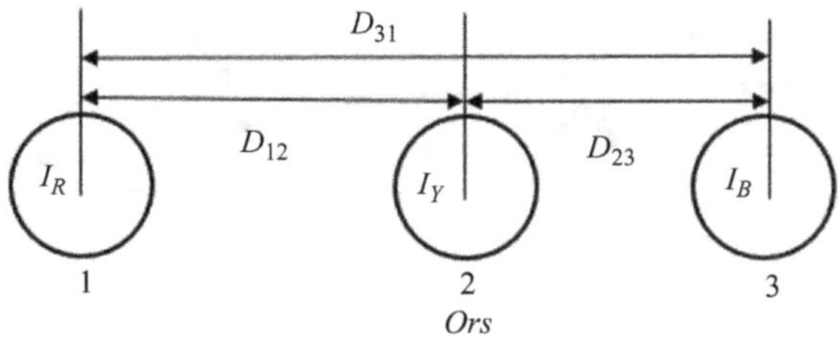

Figure 2.4 Three unsymmetrically placed conductors

$$I_R = I_Y = I_B \tag{2.34}$$

Now, as inter-distances are unequal,

$$D_{12} \neq D_{23} \neq D_{31} \tag{2.35}$$

The flux linkage corresponding to each phase will not be the same, i.e.

$$\varphi_R \neq \varphi_Y \neq \varphi_B \tag{2.36}$$

It results in

$$L_R \neq L_Y \neq L_B \tag{2.37}$$

$$\omega L_R \neq \omega L_Y \neq \omega L_B$$

$$X_R \neq X_Y \neq X_B \tag{2.38}$$

Therefore, voltage drop in each phase will not be equal, i.e.

$$IX_R \neq IX_Y \neq IX_B \tag{2.39}$$

Now, receiving-end voltage is equal to the difference of sending-end voltage and voltage drop in line. This will make the receiving-end voltage unbalanced. Thus, transmission line having three asymmetrically placed conductor will deliver unbalanced voltage even if balanced supply is fed at sending end and balanced load is connected at receiving end, which is not desirable.

2.5.6 Transposition

Transposition is alternate change of position of phase conductors. A transposed three-phase three-wire transmission line is shown in Figures 2.5 and 2.6.

The whole transmission system is divided into three parts of equal length. Suppose in the first part, positions 1, 2 and 3 are occupied by phases R, Y and B, respectively, then in the second part, positions 1, 2 and 3 are occupied by phases B, R and Y and in the third part, they are occupied by phases Y, B and R, respectively.

2.5.6.1 Advantage of transposition

In each individual part, flux linkage will not be equal, i.e. $\varphi_R \neq \varphi_Y \neq \varphi_B$, and hence, average inductance of each phase will not be equal, i.e. $L_R \neq L_Y \neq L_B$.

However, if the whole line length of the transmission line is considered, average flux linkage of each phase will be the same, i.e. $\overline{\varphi}_R = \overline{\varphi}_Y = \overline{\varphi}_B$, and hence, average

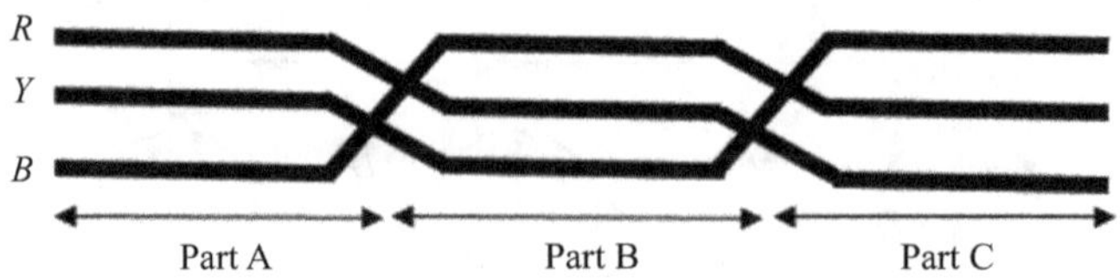

Figure 2.5 Transposed three-phase three-wire transmission line

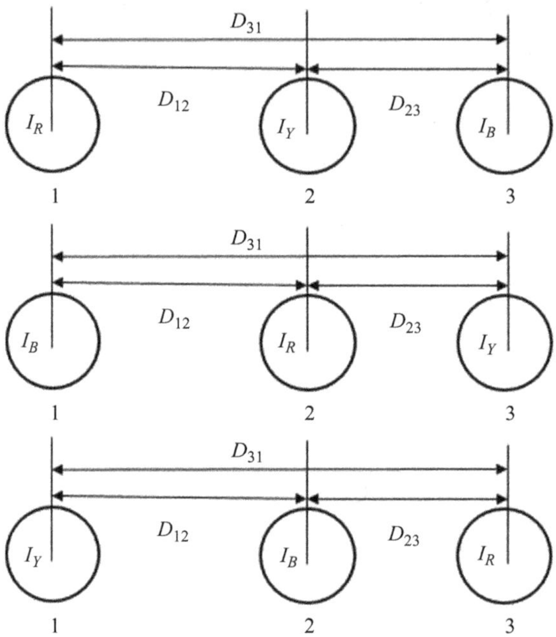

Figure 2.6 Conductor positions in a transposed three-phase three-wire transmission line

inductance of each phase will be equal, i.e. $\overline{L}_R = \overline{L}_Y = \overline{L}_B$. This will make same voltage drop for balanced supply and balanced load and the transmission system will provide balanced voltage at the receiving end helping the overall system to be balanced.

2.5.7 Determination of inductance of three-phase transmission line with three unsymmetrical but transposed wires

Consider a three-phase transmission system made of unsymmetrically placed three conductors which are transposed. Here inter-distances are not the same, i.e.

$$D_{12} \neq D_{23} \neq D_{31}$$

and the radii of the conductors are R_1, R_2 and R_3, respectively.

Let the current in R, Y and B phases be I_R, I_Y and I_B, respectively.

2.5.7.1 Flux linkage in phase *R*

Flux linkage in phase R will be created by I_R (self) and by I_Y and I_B (mutual). In the first part, i.e. in first one-third length, flux linkage in phase R can be written as

$$\varphi_{R_1} = 2 \times 10^{-7}\left(I_R \ln \frac{1}{R_1} + I_Y \ln \frac{1}{D_{12}} + I_B \ln \frac{1}{D_{31}}\right) \tag{2.40}$$

Similarly, in second part, i.e. in second one-third, flux linkage in phase R can be written as follows:

$$\varphi_{R_2} = 2 \times 10^{-7}\left(I_R \ln \frac{1}{R_1} + I_Y \ln \frac{1}{D_{23}} + I_B \ln \frac{1}{D_{12}}\right) \tag{2.41}$$

Similarly, in last part, i.e. in third portion, flux linkage in phase R can be written as follows:

$$\varphi_{R_3} = 2 \times 10^{-7}\left(I_R \ln \frac{1}{R_1} + I_Y \ln \frac{1}{D_{31}} + I_B \ln \frac{1}{D_{23}}\right) \tag{2.42}$$

Therefore, average flux linkage in phase R will be as follows:

$$\begin{aligned}
\overline{\varphi_R} &= \frac{\varphi_{R_1} + \varphi_{R_2} + \varphi_{R_3}}{3} \\
&= \frac{1}{3} \times 2 \times 10^{-7}\left(3\, I_R \ln \frac{1}{R_1} + I_Y \ln \frac{1}{D_{12}D_{23}D_{31}} + I_B \ln \frac{1}{D_{12}D_{23}D_{31}}\right) \\
&= \frac{1}{3} \times 2 \times 10^{-7}\left(3\, I_R \ln \frac{1}{R_1} - 3I_Y \ln \sqrt[3]{D_{12}D_{23}D_{31}} - 3I_B \ln \sqrt[3]{D_{12}D_{23}D_{31}}\right) \\
&= 2 \times 10^{-7}\left(I_R \ln \frac{1}{R_1} - I_Y \ln \sqrt[3]{D_{12}D_{23}D_{31}} - I_B \ln \sqrt[3]{D_{12}D_{23}D_{31}}\right) \\
&= 2 \times 10^{-7}\left\{I_R \ln \frac{1}{R_1} - (I_Y + I_B)\ln \sqrt[3]{D_{12}D_{23}D_{31}}\right\}
\end{aligned} \tag{2.43}$$

If the system is balanced,

$$I_R + I_Y + I_B = 0$$

or,

$$I_Y + I_B = -I_R$$

$$\overline{\varphi_R} = 2 \times 10^{-7}\left\{3I_R \ln \frac{1}{R_1} + I_R \ln \sqrt[3]{D_{12}D_{23}D_{31}}\right\} \tag{2.44}$$

$$\overline{\varphi_R} = 2 \times 10^{-7} I_R \ln \frac{\sqrt[3]{D_{12}D_{23}D_{31}}}{R_1} \tag{2.45}$$

Let,

$$\sqrt[3]{D_{12}D_{23}D_{31}} = D \tag{2.46}$$

$$\overline{\varphi_R} = 2 \times 10^{-7} I_R \ln \frac{D}{R_1} \tag{2.47}$$

Therefore, average inductance in R phase can be written as follows:

$$\overline{L_R} = 2 \times 10^{-7} \ln \frac{D}{R_1} \tag{2.48}$$

Similarly, average inductance in Y phase can be written as follows:

$$\overline{L_Y} = 2 \times 10^{-7} \ln \frac{D}{R_2} \tag{2.49}$$

and average inductance in B phase can be written as follows:

$$\overline{L_B} = 2 \times 10^{-7} \ln \frac{D}{R_3} \tag{2.50}$$

Average distance (D) refers to geometrical mean distance (GMD) that has been discussed in the next section.

2.5.8 Geometrical mean distance

Let us consider that conductors are denoted as 1, 2, 3, …, n as shown in Figure 2.7.

Distance between 1st and 2nd conductors $= D_{12}$
Distance between 2nd and 3rd conductors $= D_{23}$
⋮
Distance between $(n-1)$th and nth conductors $= D_{(n-1)n}$
Distance between nth and 1st conductors $= D_{n1}$

Then GMD is defined as follows:

$$GMD = \sqrt[n]{D_{12} D_{23} D_{34} \cdots D_{(n-1)n} D_{n1}} \tag{2.51}$$

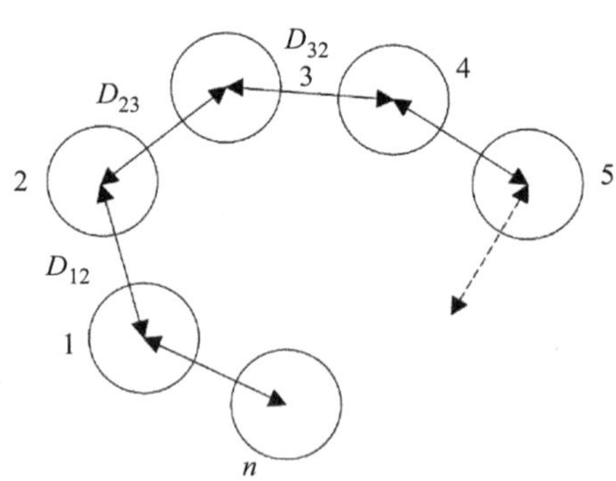

Figure 2.7 'n'-number of conductors

Therefore, for a three-wire system, GMD can be written as follows:

$$GMD = \sqrt[3]{D_{12}D_{23}D_{31}} = D \text{ (as defined in the previous expression)} \quad (2.52)$$

2.5.9 Geometrical mean radius

Let us consider that conductors are denoted as 1, 2, 3, ... n as shown in Figure 2.8.

Radius of 1st conductor $= R_1$
Radius of 2nd conductor $= R_2$
Radius of 3rd conductor $= R_3$
⋮
Radius of nth conductor $= R_n$

Then, geometrical mean radius (GMR) is defined as follows:

$$GMR = \sqrt[n]{R_1 R_2 R_3 \cdots R_{(n-1)} R_n} \quad (2.53)$$

Therefore, for a three-wire system, GMR can be written as follows:

$$GMD = \sqrt[3]{R_1 R_2 R_3} = R \quad (2.54)$$

2.5.10 Inductance in terms of GMD and GMR

In terms of GMD and GMR, the inductance of a line can be written as follows:

$$\overline{L_B} = 2 \times 10^{-7} \ln \frac{GMD}{GMR} \quad (2.55)$$

Therefore, geometrical mean distance (GMD) and geometrical mean radius (GMR) play important role in determination of line inductance.

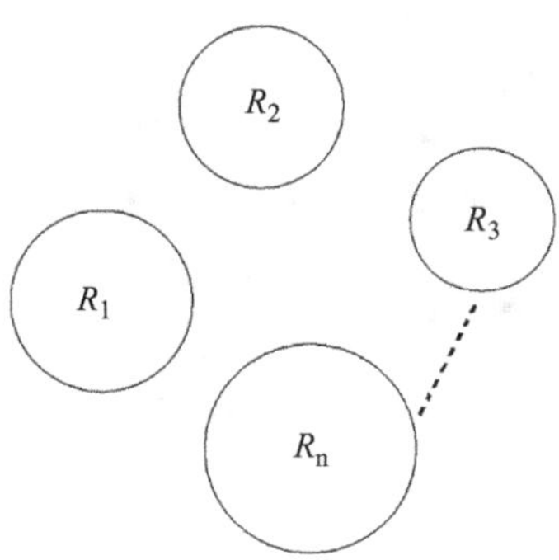

Figure 2.8 'n'-number of conductors of different radius

2.6 Skin effect

Resistance of a material where current is uniformly distributed can be determined from the following formula:

$$\text{Resistance} = \text{resistivity} \times \frac{\text{length}}{\text{cross-sectional area}} \tag{2.56}$$

In those cases, resistance is uniformly distributed, i.e. resistivity is the same at all parts of the material. But in conducting material, current is not uniformly distributed. Current is made of free electrons, which are negatively charged. So, all free electrons will try to repulse each other. Thus, electrons try to go apart from each other. Thus, electron density becomes high at periphery and low at centre. This results in high resistance at centre, which is gradually decreasing along the radius towards periphery. This is known as skin effect. It increases effective resistance and overall loss.

Non-uniformity of resistance is almost the same both in AC and DC. But in AC, other than non-uniform resistive effect, there is an effect of non-uniform flux linkage. Internal flux linkage is high at the centre and minimum at the periphery. Thus, overall impedance is high at centre and minimum at periphery. Thus, skin effect is more in AC than in DC.

Skin effect can be defined as a phenomenon by which current density becomes high near to the periphery and becomes low near to the centre.

2.6.1 Skin effect increases overall loss

Consider a conductor which is made of *n* number of small conductors as shown in Figure 2.9.

First, let us assume that there is no skin effect. The total resistance of conductor is R, i.e. if resistance is uniformly distributed, each small conductor will have resistance nR.

Let current I flow through the conductor. Therefore, current through each conductor will be (I/n) as current density is the same everywhere due to uniform resistance.

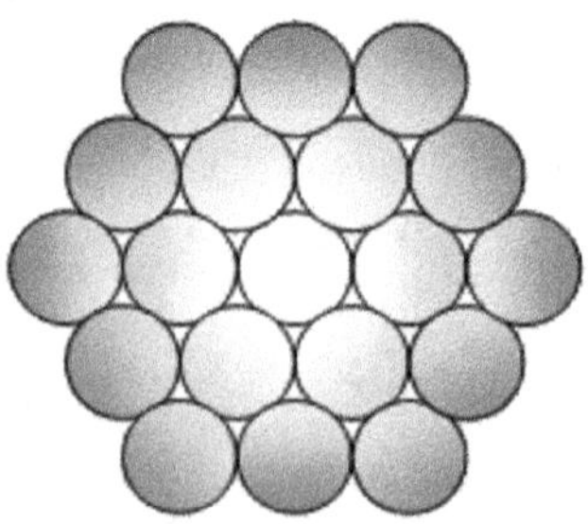

Figure 2.9 Conductor made of 'n'-number of small conductors

Therefore, loss in each small conductor will be

$$\left(\frac{I}{n}\right)^2 nR$$

Therefore, total loss considering all n conductors will be

$$n \times \left(\frac{I}{n}\right)^2 nR = I^2 R \tag{2.57}$$

Let us now assume that the same conductor is having skin effect. For simplicity, let us divide the cross section into two sections; each section consists of $n/2$ small circular conductor. Due to skin effect, current through each circle of inner part will be less than I/n, say it is $((I/n) - I')$ and the current through each circle of outer part will be greater than I/n, say it is $((I/n) + I')$.

Due to skin effect, resistance of each circle of outer part will be less than nR, say it is $(nR + \Delta R)$ and resistance of each circle of outer part will be less than nR, say it is $(nR - \Delta R)$.

Therefore,

$$\text{loss occurred in inner part} = \frac{n}{2}\left(\frac{I}{n} - I'\right)^2 (nR + \Delta R) \tag{2.58}$$

and

$$\text{loss occurred in outer part} = \frac{n}{2}\left(\frac{I}{n} + I'\right)^2 (nR - \Delta R) \tag{2.59}$$

$$\begin{aligned}\text{Therefore, total loss} &= \text{loss in inner part} + \text{loss in outer part}\\ &= \left\{\frac{n}{2}\left(\frac{I}{n} - I'\right)^2 (nR + \Delta R) + \frac{n}{2}\left(\frac{I}{n} + I'\right)^2 (nR - \Delta R)\right\}\\ &= \frac{n}{2}\left\{\left(\frac{I}{n} - I'\right)^2 (nR + \Delta R) + \left(\frac{I}{n} + I'\right)^2 (nR - \Delta R)\right\}\end{aligned} \tag{2.60}$$

As computational effect of square of change of current is higher than effect of change of resistance, for simplicity neglecting the latter effect, loss will be

$$\begin{aligned}&\frac{n}{2}\left\{\left(\frac{I}{n} - I'\right)^2 (nR) + \left(\frac{I}{n} + I'\right)^2 (nR)\right\}\\ &= \frac{n^2 R}{2}\left\{\left(\frac{I}{n} - I'\right)^2 + \left(\frac{I}{n} + I'\right)^2\right\}\\ &= I^2 R + (nI')^2 R\end{aligned} \tag{2.61}$$

The previous expression shows that loss considering skin effect is greater than loss without skin effect. Thus, skin effect increases effective resistance and hence net loss.

2.7 Proximity effect

Proximity effect refers to unequal distribution of current density over cross section of line conductors due to the effect of flux distribution of conductors that are very close to it. When two or more conductors carry electric power, due to the interaction of external flux of those conductors, current density of different parts of the cross section differs. It also depends on relative direction of flow of power. If current flows in same direction, repulsion takes place and current density decreases in nearby parts of cross section. When current flows in opposite direction, current density increases in nearby part of the cross section. This unequal distribution is known as proximity effect. It increases effective resistance of the conductor.

2.8 Method of determination of effective resistance

To determine effective resistance (R_{eff}), first power loss (P) is to be measured. Then, effective resistance may be determined as follows:

$$I^2 R_{eff} = P$$
$$\therefore R_{eff} = \frac{P}{I^2} \tag{2.62}$$

where, I is the current flowing through line conductor. This effective resistance includes the effect of both skin effect and proximity effect.

2.9 Capacitance

Overhead line conductors are separated from each other and from ground or earth by air. Earth is considered conducting material at zero potential. Thus, conductors form capacitor between each other and with earth.

2.9.1 Capacitance of a single-phase two-wire transmission line

Consider a single-phase two-wire transmission line as shown in Figure 2.10.

Let the distance between two conductors be D and their radii be R_a and R_b. The line is fed by single-phase AC supply which develops equal and opposite charges in two conductors, i.e.

$$q_a = -q_b \tag{2.63}$$

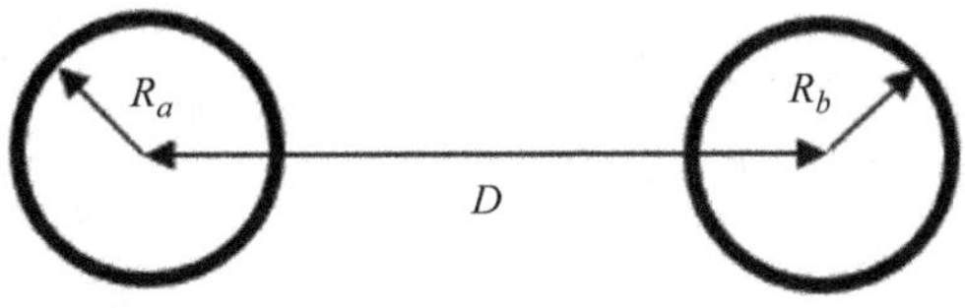

Figure 2.10 Single-phase two-wire transmission line

Voltage will be introduced due to both charges. Voltage across *a* and *b* due to charge at *a* may be written as follows:

$$V_{ab_{qa}} = \frac{q_a}{2\pi\varepsilon} \ln \frac{D}{R_a} \tag{2.64}$$

Similarly, voltage across *a* and *b* due to charge at *b* may be written as follows:

$$V_{ab_{q_b}} = \frac{q_b}{2\pi\varepsilon} \ln \frac{R_b}{D} \tag{2.65}$$

Therefore, net voltage across *a* and *b* can be written as follows:

$$\begin{aligned} V_{ab} &= V_{ab_{qa}} + V_{ab_{qb}} \\ &= \frac{q_a}{2\pi\varepsilon} \ln \frac{D}{R_a} + \frac{q_b}{2\pi\varepsilon} \ln \frac{R_b}{D} \\ &= \frac{1}{2\pi\varepsilon}\left(q_a \ln \frac{D}{R_a} + q_b \ln \frac{R_b}{D}\right) \\ &= \frac{1}{2\pi\varepsilon}\left(q_a \ln \frac{D}{R_a} - q_a \ln \frac{R_b}{D}\right) \\ &= \frac{1}{2\pi\varepsilon}\left(q_a \ln \frac{D}{R_a} + q_a \ln \frac{D}{R_b}\right) \\ &= \frac{1}{2\pi\varepsilon}\left(q_a \ln \frac{D^2}{R_a R_b}\right) \\ &= \frac{1}{2\pi\varepsilon} \times 2q_a \ln \frac{D}{\sqrt{R_a R_b}} \\ &= \frac{1}{\pi\varepsilon} \times q_a \ln \frac{D}{\sqrt{R_a R_b}} \end{aligned} \tag{2.66}$$

Let, $|q_a| = |-q_b| = q$ and $GMR = \sqrt{R_a R_b} = R$

Therefore,

$$V_{ab} = \frac{q}{\pi\varepsilon} \ln \frac{D}{R} \tag{2.67}$$

Therefore, capacitance across two conductors can be written as follows:

$$C_{ab} = \frac{q}{V_{ab}} = \frac{\pi\varepsilon}{\ln\frac{D}{R}} \tag{2.68}$$

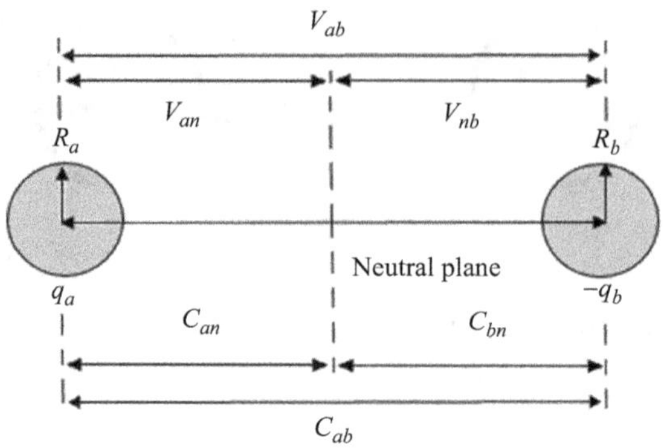

Figure 2.11 Capacitance framed by neutral plane and conductor

In between two conductors, there will be one neutral potential line, and voltage across two conductors can be written as follows:

$$V_{ab} = V_{an} + V_{nb} \tag{2.69}$$

Capacitance framed with neutral plane is shown in Figure 2.11.

Net capacitance can be written as follows:

$$C_{ab} = \frac{C_{an}}{2} = \frac{C_{nb}}{2} \tag{2.70}$$

Therefore,

$$C_{an} = C_{bn} = 2C_{ab} = \frac{2\pi\varepsilon}{\ln\frac{D}{R}} \tag{2.71}$$

2.9.2 Determination of capacitance of three-phase transmission line with three wires placed symmetrically

Consider a three-phase transmission system made of three conductors placed symmetrically as shown in Figure 2.12. Let the distance between any two conductors be D.

Voltage will be developed across a and b due to the charge placed at a and b as well as the charge placed at c.

Therefore, voltage across a and b can be written as follows:

$$\begin{aligned} V_{ab} &= V_{ab_{qa}} + V_{ab_{qb}} + V_{ab_{qc}} \\ &= \frac{q_a}{2\pi\varepsilon}\ln\frac{D}{R_a} + \frac{q_b}{2\pi\varepsilon}\ln\frac{R_b}{D} + \frac{q_c}{2\pi\varepsilon}\ln\frac{D}{D} \\ &= \frac{1}{2\pi\varepsilon}\left(q_a\ln\frac{D}{R_a} + q_b\ln\frac{R_b}{D}\right) \end{aligned} \tag{2.72}$$

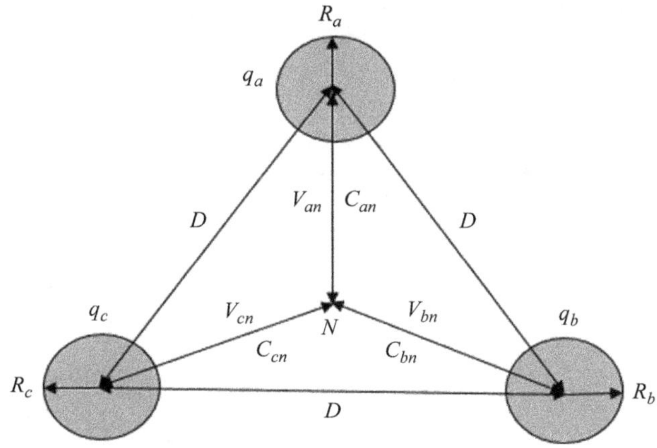

Figure 2.12 Three-phase transmission line with three symmetrically placed conductors

Similarly, voltage across a and c be can be written as follows:

$$\begin{aligned}V_{ac} &= V_{ac_{qa}} + V_{ac_{qb}} + V_{ac_{qc}} \\ &= \frac{q_a}{2\pi\varepsilon}\ln\frac{D}{R_a} + \frac{q_b}{2\pi\varepsilon}\ln\frac{D}{D} + \frac{q_c}{2\pi\varepsilon}\ln\frac{R_c}{D} \\ &= \frac{1}{2\pi\varepsilon}\left(q_a\ln\frac{D}{R_a} + q_c\ln\frac{R_c}{D}\right)\end{aligned} \tag{2.73}$$

Therefore,

$$V_{ab} + V_{ac} = \frac{1}{2\pi\varepsilon}\left(q_a\ln\frac{D}{R_a} + q_b\ln\frac{R_b}{D}\right) + \frac{1}{2\pi\varepsilon}\left(q_a\ln\frac{D}{R_a} + q_c\ln\frac{R_c}{D}\right) \tag{2.74}$$

Let the radius of each conductor be R. So,

$$V_{ab} + V_{ac} = \frac{1}{2\pi\varepsilon}\left\{2q_a\ln\frac{D}{R} + (q_b + q_c)\ln\frac{R}{D}\right\} \tag{2.75}$$

Now for balanced system,

$$q_a + q_b + q_c = 0 \tag{2.76}$$

Therefore,

$$V_{ab} + V_{ac} = \frac{1}{2\pi\varepsilon}\left\{2q_a\ln\frac{D}{R} - q_a\ln\frac{R}{D}\right\} \tag{2.77}$$

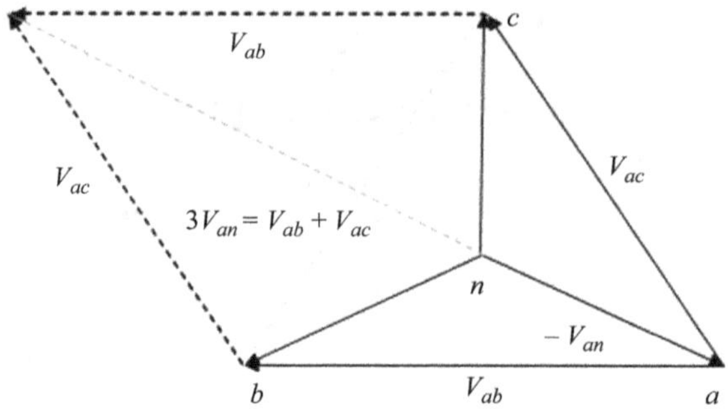

Figure 2.13 Phasor diagram corresponding to voltages

$$V_{ab} + V_{ac} = \frac{1}{2\pi\varepsilon} 3q_a \ln \frac{D}{R} \tag{2.78}$$

Now from the phasor diagram as shown in Figure 2.13,

$$V_{ab} + V_{ac} = 3V_{an} \tag{2.79}$$

Therefore,

$$\begin{aligned} 3V_{an} &= V_{ab} + V_{ac} \\ &= \frac{1}{2\pi\varepsilon} 3q_a \ln \frac{D}{R} \end{aligned} \tag{2.80}$$

$$V_{an} = \frac{1}{2\pi\varepsilon} q_a \ln \frac{D}{R} = V_{bn} = V_{cn} \tag{2.81}$$

Therefore, capacitance across neutral to phase a can be written as follows:

$$C_{an} = \frac{q}{V_{an}} = \frac{2\pi\varepsilon}{\ln\frac{D}{R}} = C_{bn} = C_{cn} \tag{2.82}$$

2.9.3 Determination of capacitance of three-phase transmission line with three unsymmetrical but transposed wires

Consider a three-phase transmission system made of unsymmetrically placed three conductors which are transposed as shown in Figures 2.14 and 2.15. Here, inter-distances are not same, i.e.

$$D_{12} \neq D_{23} \neq D_{31}$$

Let the radii of the conductors be R_a, R_b and R_c, respectively, and the charges at in phases a, b and c phases be q_a, q_b and q_c, respectively.

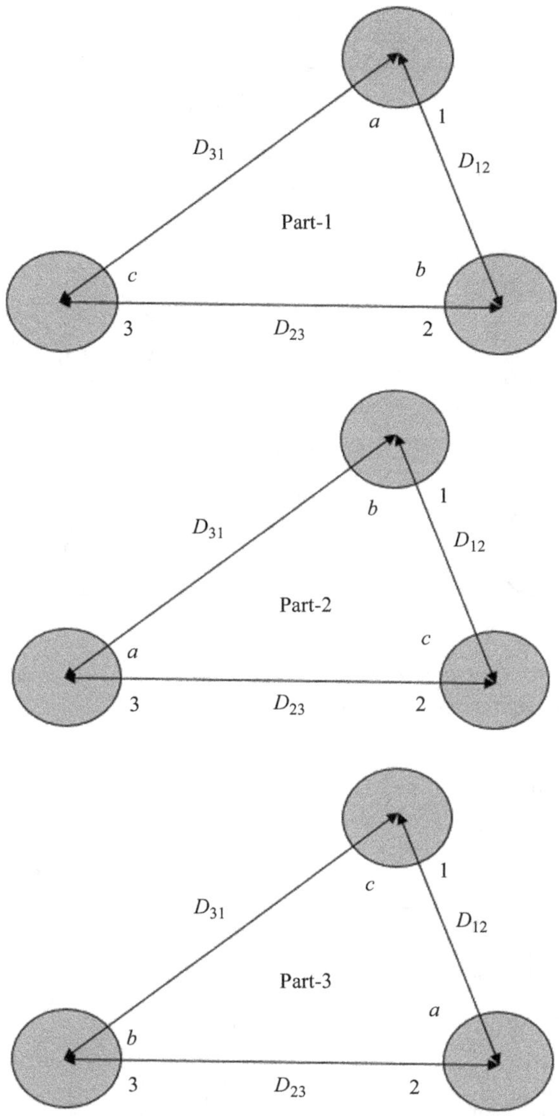

Figure 2.14 Three-phase line with conductors placed in unsymmetrical but transposed manner

The whole transmission system is divided into three parts of equal length. Suppose in the first part, positions 1, 2 and 3 are occupied by phases *a*, *b* and *c*, respectively, then in the second part, positions 1, 2 and 3 are occupied by phases *c*, *a* and *b* and in the third part, they are occupied by phases *b*, *c* and *a*, respectively.

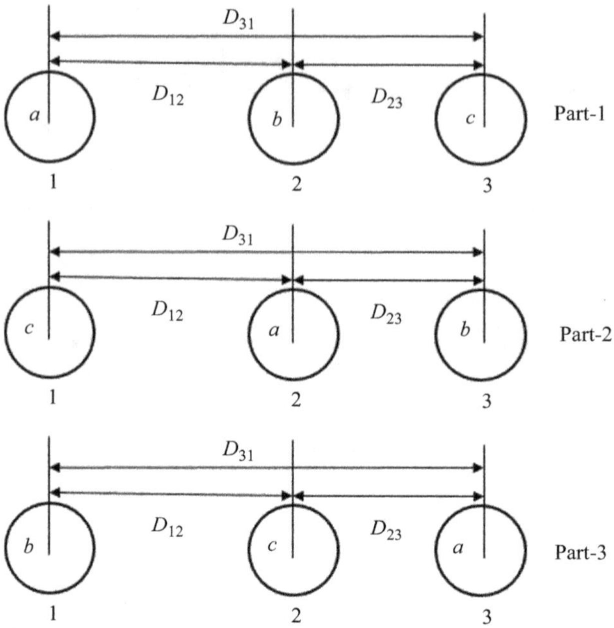

Figure 2.15 Conductor positions in a three-phase line with conductors placed in unsymmetrical but transposed manner

In the first part,

$$V_{ab1} = V_{ab_{qa}} + V_{ab_{q_b}} + V_{ab_{qc}} \\ = \frac{q_a}{2\pi\varepsilon}\ln\frac{D_{12}}{R_a} + \frac{q_b}{2\pi\varepsilon}\ln\frac{R_b}{D_{12}} + \frac{q_c}{2\pi\varepsilon}\ln\frac{D_{23}}{D_{31}} \tag{2.83}$$

Similarly, in the second part,

$$V_{ab2} = V_{ab_{qa}} + V_{ab_{qb}} + V_{ab_{qc}} \\ = \frac{q_a}{2\pi\varepsilon}\ln\frac{D_{23}}{R_a} + \frac{q_b}{2\pi\varepsilon}\ln\frac{R_b}{D_{23}} + \frac{q_c}{2\pi\varepsilon}\ln\frac{D_{31}}{D_{12}} \tag{2.84}$$

Similarly, in the third part,

$$V_{ab3} = V_{ab_{qa}} + V_{ab_{q_b}} + V_{ab_{qc}} \\ = \frac{q_a}{2\pi\varepsilon}\ln\frac{D_{31}}{R_a} + \frac{q_b}{2\pi\varepsilon}\ln\frac{R_b}{D_{31}} + \frac{q_c}{2\pi\varepsilon}\ln\frac{D_{12}}{D_{23}} \tag{2.85}$$

Therefore, average voltage across a and b will be

$$\begin{aligned} V_{ab} &= \frac{V_{ab1} + V_{ab2} + V_{ab3}}{3} \\ &= \frac{q_a}{2\pi\varepsilon} \ln \frac{D_{12}D_{23}D_{31}}{R_a{}^3} + \frac{q_b}{2\pi\varepsilon} \ln \frac{R_b{}^3}{D_{12}D_{23}D_{31}} + \frac{q_c}{2\pi\varepsilon} \ln \frac{D_{12}D_{23}D_{31}}{D_{12}D_{23}D_{31}} \\ &= \frac{q_a}{2\pi\varepsilon} \ln \frac{D_{12}D_{23}D_{31}}{R_a{}^3} + \frac{q_b}{2\pi\varepsilon} \ln \frac{R_b{}^3}{D_{12}D_{23}D_{31}} \\ &= \frac{3q_a}{2\pi\varepsilon} \ln \frac{\sqrt[3]{D_{12}D_{23}D_{31}}}{R_a} + \frac{3q_b}{2\pi\varepsilon} \ln \frac{R_b}{\sqrt[3]{D_{12}D_{23}D_{31}}} \\ &= \frac{3q_a}{2\pi\varepsilon} \ln \frac{D}{R_a} + \frac{3q_b}{2\pi\varepsilon} \ln \frac{R_b}{D} \\ &= \frac{3}{2\pi\varepsilon} \left(q_a \ln \frac{D}{R_a} + q_b \ln \frac{R_b}{D} \right) \end{aligned} \tag{2.86}$$

$$GMD = D = \sqrt[3]{D_{12}D_{23}D_{31}} \tag{2.87}$$

Similarly,

$$V_{ac} = \frac{3}{2\pi\varepsilon} \left(q_a \ln \frac{D}{R_a} + q_c \ln \frac{R_c}{D} \right) \tag{2.88}$$

Therefore,

$$V_{ab} + V_{ac} = \frac{1}{2\pi\varepsilon} \left(q_a \ln \frac{D}{R_a} + q_b \ln \frac{R_b}{D} \right) + \frac{1}{2\pi\varepsilon} \left(q_a \ln \frac{D}{R_a} + q_c \ln \frac{R_c}{D} \right) \tag{2.89}$$

Let the radius of each conductor be R, so

$$V_{ab} + V_{ac} = \frac{1}{2\pi\varepsilon} \left\{ 2q_a \ln \frac{D}{R} + (q_b + q_c) \ln \frac{R}{D} \right\} \tag{2.90}$$

Now for balanced system,

$$q_a + q_b + q_c = 0 \tag{2.91}$$

Therefore,

$$V_{ab} + V_{ac} = \frac{1}{2\pi\varepsilon} \left\{ 2q_a \ln \frac{D}{R} - q_a \ln \frac{R}{D} \right\} \tag{2.92}$$

$$V_{ab} + V_{ac} = \frac{1}{2\pi\varepsilon} 3q_a \ln \frac{D}{R} \tag{2.93}$$

Now, from the phasor diagram (Figure 2.13)

$$V_{ab} + V_{ac} = 3V_{an} \tag{2.94}$$

Therefore,

$$3V_{an} = V_{ab} + V_{ac}$$
$$= \frac{1}{2\pi\varepsilon} 3q_a \ln \frac{D}{R} \tag{2.95}$$

$$V_{an} = \frac{1}{2\pi\varepsilon} q_a \ln \frac{D}{R} \tag{2.96}$$

Therefore, capacitance across neutral to phase *a* can be written as follows:

$$C_{an} = \frac{q}{V_{an}} = \frac{2\pi\varepsilon}{\ln \frac{D}{R}} \tag{2.97}$$

2.10 Sequence impedance

Impedance in the path of the sequence component of current is known as sequence impedance. There are three sequence impedances:

1. *Positive-sequence impedance:* Impedance in the path of positive-sequence current is known as positive-sequence impedance.
2. *Negative-sequence impedance:* Impedance in the path of negative-sequence current is known as negative-sequence impedance.
3. *Zero-sequence impedance:* Impedance in the path of zero-sequence current is known as zero-sequence impedance.

Equivalent sequence networks consisting of zero, positive and negative-sequence components are shown in Figure 2.16. Among these networks, only positive-sequence network has voltage source having magnitude equal to the voltage at balanced condition. Negative and zero-sequence networks do not have any source; power of these networks depends on the positive-sequence network. Z_1, Z_2 and Z_0 are positive-sequence, negative-sequence and zero-sequence impedances, respectively.

For linear, symmetrical and static system, positive and negative sequences are equal in magnitude, and they are equal to the impedance seen at balanced condition provided the supply is balanced. However, in rotating machines, positive- and negative-sequence impedances are usually different in magnitude. Zero-sequence impedance is usually different from the other two sequence impedances as the zero-sequence path is different from the positive and negative-sequence paths.

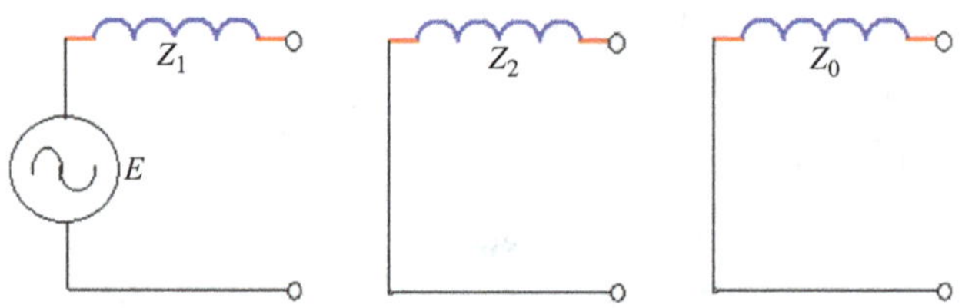

Figure 2.16 Equivalent sequence network

Unbalanced voltages and current can be expressed in terms of positive-sequence, negative-sequence and zero-sequence voltage and current components. Rotation of negative-sequence components are opposite to positive-sequence components. Zero-sequence voltage and currents do not have rotational behaviour. In three-phase system, separately, each of positive, negative and zero-sequence components are balanced, but, in combination, they form unbalance condition of resultant components.

During unbalance of a system, zero-sequence current or voltage may or may not flow in the system. Some special cases regarding zero-sequence current are given as follows:

- In balanced system, zero-sequence component is absent.
- In unbalanced star-connected four-wire system, zero-sequence component flows though neutral wire. It is equal to the sum of the three-line currents.
- In unbalanced ungrounded star-connected three-wire system, there is no zero-sequence current as there is no return or neutral path for zero-sequence current.
- In unbalanced delta connected system, zero-sequence current circulates through the mesh. In line current, it is absent.

During unbalance or unsymmetrical fault, negative component of current occurs. It changes the resultant current. Negative-sequence components may be detected by negative-sequence relay consisting of negative-sequence filter.

2.11 Short transmission line

Short transmission line (STL) has a very considerable resistance, inductance and very small capacitance, which is normally neglected in modelling and analysis. Transmission lines whose length is less than 50 km are normally considered as STL. However, line length demarcation is less important. More important is that amount of capacitance is negligible. Mainly, STLs are used in distribution side and are operated with low voltage (LV) or low tension (LT).

2.11.1 Model of short transmission line

For modelling of STL, parameters are represented by lumped parameters. Equivalent lumped resistance (R) is connected in series with lumped inductance (L) as shown in Figure 2.17.

2.11.2 Regulation of short transmission line

Regulation of a transmission line is defined as follows:

$$\text{Regulation} = \frac{\text{receiving end voltage at no-load} - \text{receiving end voltage at full-load}}{\text{receiving end voltage at full-load}} \tag{2.98}$$

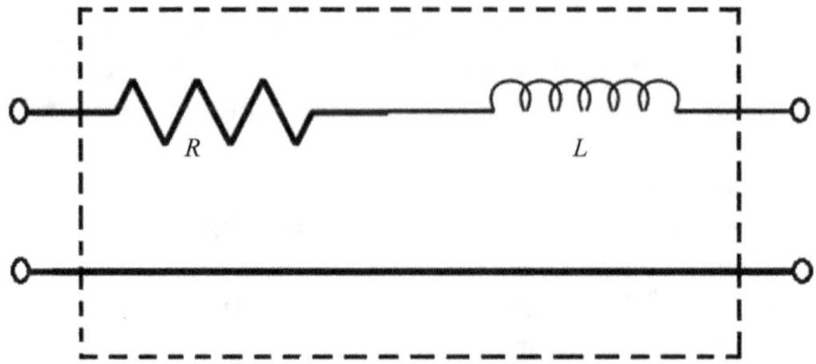

Figure 2.17 Short transmission line

Now for STL,

Receiving end voltage at no-load = sending end voltage at full-load

Therefore,

$$\text{Regulation} = \frac{\text{sending end voltage at full-load} - \text{receiving end voltage at full-load}}{\text{receiving end voltage at full-load}} \tag{2.99}$$

In the model shown in Figure 2.17, say,

V_S = sending end voltage

V_R = receiving end voltage

I_S = sending end current

I_R = receiving end current

In STL,

$$I_S = I_R$$

Let,

$$I_S = I_R = I$$

R = line resistance
L = line inductance

STL is made of resistance and inductive reactance and is connected with inductive load, and the model will be an inductive network. Current flowing through line is inductive, i.e. lagging in nature, and the drop will be an inductive drop. The phasor is shown in Figure 2.18.

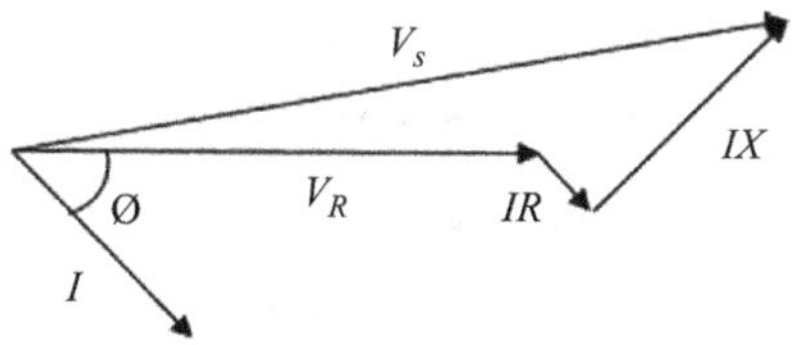

Figure 2.18 Phasor of short transmission line

In phasor diagram, *IR* is the drop across line resistance and is parallel to *I*, and *IX* is the drop across inductance and is perpendicular to *I* or *IR*. Sending-end voltage (V_S) is the vector sum of receiving-end voltage (V_R), drop across resistance (*IR*), and drop across reactance (*IX*). Phasor shows that sending-end voltage is greater in magnitude than receiving-end voltage. From geometry of the phasor, sending-end voltage can be written as follows:

$$V_S^2 = (V_S \cos \varnothing + IR)^2 + (V_R \sin \varnothing + IX)^2$$

$$V_S = \sqrt{(V_S \cos \varnothing + IR)^2 + (V_S \sin \varnothing + IX)^2}$$

$$V_S = \sqrt{V_S^2 + 2V_R IR \cos \varnothing + 2V_R IX \sin \varnothing + (IR)^2 (IX)^2}$$

$$V_S \cong \sqrt{V_S^2 + 2V_R IR \cos \varnothing + 2V_R IX \sin \varnothing}$$

$$V_S \cong V_R \left(1 + \frac{2IR \cos \varnothing + 2IX \sin \varnothing}{V_R}\right)^{1/2} \tag{2.100}$$

$$V_S \cong V_R \left(1 + \frac{IR \cos \varnothing + IX \sin \varnothing}{V_R}\right)$$

$$\therefore \text{Regulation } \% = \frac{V_S - V_R}{V_S} \times 100\%$$

$$\cong \frac{IR \cos \varnothing + IX \sin \varnothing}{V_S} \times 100\%$$

2.11.3 Transmission (ABCD) parameters

Two-port network model refers to linear, bilateral and passive network having two ports. As transmission line is made of resistance, inductance and capacitance that are linear, bilateral passive in nature, transmission line can be modelled with two-port network model. Voltage and current of sending end of two-port network can be expressed in terms of voltage and current of receiving end as follows:

$$\begin{bmatrix} V_S \\ I_R \end{bmatrix} = \begin{bmatrix} A & B \\ C & D \end{bmatrix} \begin{bmatrix} I_S \\ V_R \end{bmatrix}$$

$\begin{bmatrix} A & B \\ C & D \end{bmatrix}$ is called simply ABCD parameters or transmission parameters.

A is called reverse voltage gain. It is defined as the ratio of the sending-end voltage and receiving-end voltage when receiving end is open. Mathematically,

$$A = V_S/V_R \text{ when } I_R = 0, \text{i.e. output is open.}$$

B is called transfer impedance. It is defined as the ratio of the sending-end voltage and receiving-end current when receiving end is shorted. Mathematically,

$$B = V_S/I_R \text{ when } V_R = 0, \text{i.e. output is shorted.}$$

C is called transfer admittance. It is defined as the ratio of the sending-end current and receiving-end voltage when receiving end is open. Mathematically,

$$C = I_S/V_R \text{ when } I_R = 0, \text{i.e. output is open.}$$

D is called reverse current gain. It is defined as the ratio of the sending-end current and receiving-end current when sending end is shorted. Mathematically,

$$D = I_S/I_R \text{ when } V_S = 0, \text{i.e. sending end is open.}$$

In power system, A, B, C and D parameters are commonly known as transmission parameters. In this approach, input variables are expresses in terms of output variables. This approach is very much useful in steady-state performance analysis of transmission line. Conditions of symmetry and reciprocity have been presented in Table 2.1.

2.11.4 *Transmission parameters of short transmission line*

From the model using KVL and KCL,

$$V_S = V_R + ZI_R = 1 \times V_R + (-Z)(-I_R) \tag{2.101}$$

$$I_S = -I_R = 0 \times V_R + (1)(-I_R) \tag{2.102}$$

Therefore,

$$\begin{pmatrix} V_S \\ I_S \end{pmatrix} = \begin{pmatrix} 1 & -Z \\ 0 & 1 \end{pmatrix} \begin{pmatrix} V_R \\ -I_R \end{pmatrix} \tag{2.103}$$

Therefore, transmission parameters can be written as follows:

$$\begin{pmatrix} A & B \\ C & D \end{pmatrix} = \begin{pmatrix} 1 & -Z \\ 0 & 1 \end{pmatrix} \tag{2.104}$$

Table 2.1 Condition of symmetry and reciprocity

Parameters	Condition of symmetry	Condition of reciprocity
$\begin{bmatrix} A & B \\ C & D \end{bmatrix}$	$A = B$	$AD - BC = 1$

2.11.5 Symmetry and reciprocity

Here, from (2.104), $A=D=1$, therefore, any STL is symmetrical in nature.
Also, from (2.104), $AD-BC=1$, therefore, STL is reciprocal in nature.

2.12 Medium transmission line

Medium transmission line (MTL) has considerable resistance and inductance and small capacitance. Transmission line whose length is greater than 50 km and less than 100/150 km is normally considered as MTL. Mainly MTLs are used in transmission section of power system and are operated at high voltage (HV) or high tension (HT).

2.12.1 Model of medium transmission line

Medium transmission line (MTL) is modelled by equivalent T network or Pi network. Line resistance and inductance are connected in series. Line capacitance makes the shunt path. In modelling, all parameters are assumed lumped. Admittance contributed by line capacitance is placed in the mid position of the T model (Figure 2.19) and impedance made by line resistance, and inductance is divided into two equal parts forming two arms of the T model. In Pi model, impedance made by line inductance and resistance is placed at the mid position and admittance contributed by line capacitance is divided in two equal parallel parts (twice of line capacitance in magnitude) and makes the two arms of Pi model.

2.12.2 Transmission parameters of medium transmission line

In the T model as shown in Figure 2.19, let the voltage across the capacitor (*C*) is *E*. Now, applying KVL and KCL we get,

$$V_S = I_S \times \frac{Z}{2} + E \tag{2.105}$$

$$I_S = YE + I_R \tag{2.106}$$

Now,

$$E = I_R \times \frac{Z}{2} + V_R \tag{2.107}$$

Therefore,

$$I_S = Y\left(I_R \times \frac{Z}{2} + V_R\right) + I_R \tag{2.108}$$

$$V_S = \left\{Y\left(I_R \times \frac{Z}{2} + V_R\right) + I_R\right\} \times \frac{Z}{2} + \left(I_R \times \frac{Z}{2} + V_R\right) \tag{2.109}$$

$$V_S = \left(1 + \frac{YZ}{2}\right)V_R + Z\left(1 + \frac{YZ}{4}\right)I_R \tag{2.110}$$

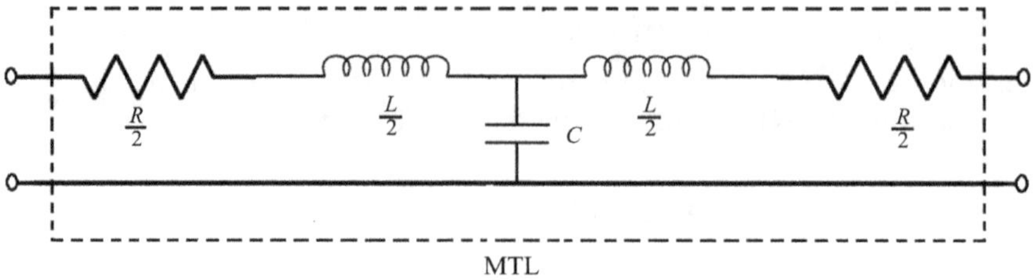

Figure 2.19 Models of medium transmission line

Therefore,

$$\begin{pmatrix} V_S \\ I_S \end{pmatrix} = \begin{pmatrix} \left(1+\frac{YZ}{2}\right) & Z\left(1+\frac{YZ}{4}\right) \\ Y & \left(1+\frac{YZ}{2}\right) \end{pmatrix} \begin{pmatrix} V_R \\ I_R \end{pmatrix} \tag{2.111}$$

$$\begin{pmatrix} A & B \\ C & D \end{pmatrix} = \begin{pmatrix} \left(1+\frac{YZ}{2}\right) & Z\left(1+\frac{YZ}{4}\right) \\ Y & \left(1+\frac{YZ}{2}\right) \end{pmatrix} \tag{2.112}$$

2.12.3 Symmetry and reciprocity

Here, $A = D = (1 + (YZ/2))$, therefore, any MTL is symmetrical in nature. Also, $AD-BC=1$, therefore, MTL is reciprocal in nature.

2.13 Long transmission line

In long transmission line (LTL), both line inductance and capacitance are very high. Resistance is small with respect to inductance and capacitance and hence is often neglected in modelling and analysis. Normally transmission lines of length greater than 100 km (or 150 km) are treated as LTLs. LTLs are used to transmit bulk amount of power and are kept at HV or extra high voltage (EHV).

2.13.1 Model of long transmission line

In modelling of LTL, all parameters are assumed distributed in nature. Thus, lumped equivalent parameters will not be applicable. Line inductance is connected along the length in series and line capacitance makes the shunt path throughout the line length. A typical model of LTL is shown in Figure 2.20.

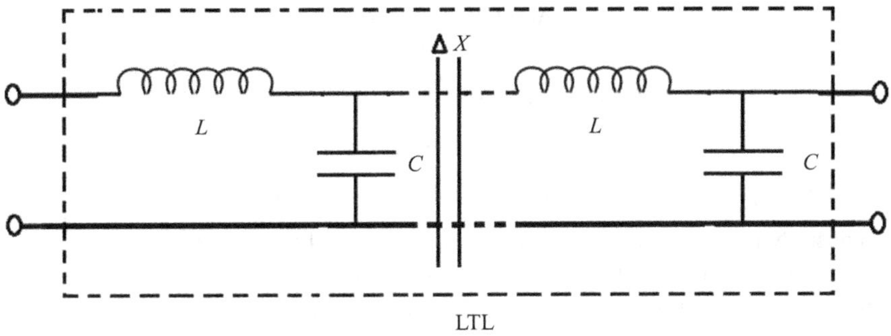

Figure 2.20 Model of long transmission line

2.13.2 Transmission parameters of long transmission line

Consider a small section of length Δx in the line where V is the receiving-end voltage and I is the receiving-end current. Also consider,

Z is the line impedance per unit length
Y is the line admittance per unit length

Therefore,

$Z\Delta x$ is the line impedance for Δx length
$Y\Delta x$ is the line admittance for Δx length

Due to the drop across line length, voltage in the sending end will be little higher than receiving end. Let it be $V + \Delta V$.

Similarly, due to shunt current flowing through admittance, sending-end current will be higher than receiving-end current; let it be $I + \Delta I$.

Now, the drop ΔV can be expressed in terms of current as follows:

$$\Delta V = IZ\Delta x \tag{2.113}$$

$$\text{or, } \frac{\Delta V}{\Delta x} = IZ \tag{2.114}$$

Now, for Δx tends to zero or very small, the previous expression will be

$$\frac{dV}{dx} = IZ \tag{2.115}$$

Similarly, the shunt current ΔI can be expressed in terms of voltage as follows:

$$\Delta I = VY\Delta x \tag{2.116}$$

$$\text{or, } \frac{\Delta I}{\Delta x} = VY \tag{2.117}$$

Now, for Δx tends to zero or very small, the previous expression will be

$$\frac{dI}{dx} = VY \tag{2.118}$$

By differentiating both equations with respect to x,

$$\frac{d^2V}{dx^2} = Z\frac{dI}{dx} = YZV \tag{2.119}$$

$$\frac{d^2I}{dx^2} = Y\frac{dV}{dx} = YZI \tag{2.120}$$

These two equations are second-order equation of voltage and current for LTL. These equations indicate that the voltage and current are sinusoidal with respect to length. Therefore, the solution can be expressed as follows:

$$V = A_1 e^{\sqrt{YZ}x} + A_2 e^{-\sqrt{YZ}x} \tag{2.121}$$

Therefore,

$$\frac{dV}{dx} = \sqrt{YZ}\left(A_1 e^{\sqrt{YZ}x} - A_2 e^{-\sqrt{YZ}x}\right) \tag{2.122}$$

Therefore,

$$IZ = \sqrt{YZ}\left(A_1 e^{\sqrt{YZ}x} - A_2 e^{-\sqrt{YZ}x}\right) \tag{2.123}$$

$$I = \sqrt{\frac{Y}{Z}}\left(A_1 e^{\sqrt{YZ}x} - A_2 e^{-\sqrt{YZ}x}\right) \tag{2.124}$$

Let,

$$\sqrt{YZ} = \gamma \text{ and } Z_C = \sqrt{\frac{Z}{Y}}$$

Therefore,

$$V = A_1 e^{\gamma x} + A_2 e^{-\gamma x} \tag{2.125}$$

$$I = \frac{1}{Z_C}\left(A_1 e^{\gamma x} - A_2 e^{-\gamma x}\right) \tag{2.126}$$

Let us impose following two boundary conditions:

i) Distance is measured from receiving end, and at receiving end $x=0$, voltage is $V = V_R$ and current is $I = I_R$.
ii) At sending end, line length is x, and voltage is V_S and current is I_S.

Applying those boundary conditions,

$$V = A_1 + A_2 \tag{2.127}$$

$$I = \frac{1}{Z_C}(A_1 - A_2) \tag{2.128}$$

Therefore,

$$A_1 = \frac{V_R + I_R Z_C}{2} \tag{2.129}$$

$$A_2 = \frac{V_R - I_R Z_C}{2} \tag{2.130}$$

Therefore, at length x, voltage and current can be written as follows:

$$\begin{aligned} V_S &= \frac{V_R + I_R Z_C}{2} e^{\gamma x} + \frac{V_R - I_R Z_C}{2} e^{-\gamma x} \\ &= V_R \frac{e^{\gamma x} + e^{-\gamma x}}{2} + I_R Z_C \frac{e^{\gamma x} + e^{-\gamma x}}{2} \\ &= \cosh \gamma x \, V_R + Z_C \sinh \gamma x \, I_R \end{aligned} \tag{2.131}$$

$$\begin{aligned} I_S &= \frac{1}{Z_C} \left(\frac{V_R + I_R Z_C}{2} e^{\gamma x} - \frac{V_R - I_R Z_C}{2} e^{-\gamma x} \right) \\ &= \frac{1}{Z_C} V_R \frac{e^{\gamma x} - e^{-\gamma x}}{2} + I_R \frac{e^{\gamma x} + e^{-\gamma x}}{2} \\ &= \frac{1}{Z_C} \sinh \gamma x \, V_R + \cosh \gamma x \, I_R \end{aligned} \tag{2.132}$$

$$\therefore \begin{pmatrix} V_S \\ I_S \end{pmatrix} = \begin{pmatrix} \cosh \gamma x & Z_C \sinh \gamma x \\ \frac{1}{Z_C} \sinh \gamma x & \cosh \gamma x \end{pmatrix} \begin{pmatrix} V_R \\ I_R \end{pmatrix} \tag{2.133}$$

Transmission line parameters are

$$\begin{pmatrix} A & B \\ C & D \end{pmatrix} = \begin{pmatrix} \cosh \gamma x & Z_C \sinh \gamma x \\ \frac{1}{Z_C} \sinh \gamma x & \cosh \gamma x \end{pmatrix} \tag{2.134}$$

2.13.3 Symmetry and reciprocity

Here, $A = D = \cosh \gamma x$, therefore, any LTL is symmetrical in nature. Also, $AD - BC = 1$, therefore, LTL is reciprocal in nature.

2.13.4 Characteristic impedance

Characteristics impedance refers to a parameter which carries characteristic information of a power line. It is derived from maximum impedance and minimum impedance either from sending end or from receiving end. Characteristics impedance is given as the square root of open-circuit impedance (maximum impedance) and short circuit impedance (minimum impedance) of any side. From the previous equation, open-circuit impedance measured at sending end when receiving end is open can be written as follows:

$$\begin{aligned} Z_{OC} &= \frac{A}{C} \\ &= \frac{\cosh \gamma x}{\frac{1}{Z_C} \sinh \gamma x} \\ &= \frac{Z_C \cosh \gamma x}{\sinh \gamma x} \end{aligned} \tag{2.135}$$

Short-circuit impedance measured at sending end when receiving end is short can be written as follows:

$$\begin{aligned} Z_{SH} &= \frac{B}{D} \\ &= \frac{Z_C \sinh \gamma x}{\cosh \gamma x} \end{aligned} \tag{2.136}$$

Therefore, characteristic impedance can be written as follows:

$$\begin{aligned} \sqrt{Z_{OC} Z_{SH}} &= \sqrt{\frac{AB}{CD}} \\ &= \sqrt{\frac{Z_C \cosh \gamma x}{\sinh \gamma x} \times \frac{Z_C \sinh \gamma x}{\cosh \gamma x}} \\ &= Z_C \end{aligned} \tag{2.137}$$

Thus, Z_C is the characteristic impedance of an LTL which is given by $Z_C = \sqrt{Z/Y}$.

2.13.5 Propagation constant

Wave velocity depends on propagation constant (γ), and it is expressed as the square root of the product of the shunt admittance per unit length and series impedance per unit length of the line. Mathematically, it may be written as follows:

$$\gamma = \sqrt{YZ}$$

2.13.6 Image impedance or surge impedance

The transmission line delivers maximum power to the load when complex conjugate of load impedance equals the characteristic impedance. During this condition, at the receiving end, impedances of both sides are equal. Thus, characteristic impedance may be considered as image of the load impedance and vice versa. At this condition, characteristic impedance is called image impedance. Image impedance is also called surge impedance.

2.13.7 Image impedance loading

When characteristic impedance matches with load impedance, i.e. load impedance is complex conjugate of characteristic impedance, the transmission line will transfer maximum power to the load. The maximum power delivered during this condition is known as image impedance loading (IIL). Mathematically,

$$\text{Image impedance loading(IIL)} = V_R I_R = \frac{V_R^2}{Z_C} \tag{2.138}$$

It is also known as surge impedance loading.

2.13.8 Wave propagation

Propagation of electromagnetic wave depends on the strength of electric field and magnetic field governed by the voltage and the current of the line. Behavior of propagation wave is analyzed by velocity, propagation constant and characteristic impedance of the line.

Propagation constant (γ) is mathematically expressed as:

$$\begin{aligned} \gamma &= \sqrt{YZ} \\ &= \sqrt{(r + j\omega L)(g + j\omega C)} \end{aligned} \tag{2.139}$$

Propagation constant can also be expressed in complex mathematical form as follows:

$$\gamma = \alpha + j\beta$$

Real part (α) is known as attenuation constant and imaginary part (β) is known as phase constant.

Propagation constant in Lossless line

For lossless line, real part of impedance (r) and real part of admittance (g) are zero. Mathematically,

$$\begin{aligned} r &= 0 \\ g &= 0 \end{aligned}$$

Therefore, propagation constant consisting of only imaginary part can be written as follows:

$$\begin{aligned} \gamma &= j\omega\sqrt{LC} \\ \therefore \gamma &= j\beta (\text{where } \beta = \omega\sqrt{LC}) \end{aligned}$$

Velocity of propagation wave in lossless line

In lossless line, velocity of propagation wave is given as follows:

$$
\begin{aligned}
v &= \frac{\omega}{\beta} \\
&= \frac{1}{\sqrt{LC}} \\
&= \frac{1}{\sqrt{\frac{\mu_0}{2\pi}\ln\frac{D}{r} \times \frac{2\pi\varepsilon_0}{\ln\frac{D}{r}}}} \\
&= \frac{1}{\sqrt{\mu_0\varepsilon_0}}
\end{aligned}
\tag{2.140}
$$

Wavelength of propagation wave in lossless line

In lossless line, wavelength (λ) of propagation wave can be expressed as follows:

$$
\begin{aligned}
\lambda &= \frac{2\pi}{\beta} \\
&= \frac{2\pi}{\omega\sqrt{LC}} \\
&= \frac{1}{f\sqrt{LC}} (\because \omega = 2\pi f) \\
&= \frac{1}{f\sqrt{\mu_0\varepsilon_0}}
\end{aligned}
\tag{2.141}
$$

Relation between velocity and wavelength of propagation wave in lossless line

Relation between velocity and wavelength of propagation wave in lossless line can be expressed as follows:

$$
\begin{aligned}
\lambda &= \frac{1}{f\sqrt{LC}} \\
&= \frac{v}{f}\left(\because v = \frac{1}{\sqrt{LC}}\right)
\end{aligned}
\tag{2.142}
$$

2.14 Comparison with AC overhead lines

2.14.1 AC lines versus DC lines

Major advantages and limitations of AC and DC lines are as follows:

Advantages of AC lines

- I^2R loss may be reduced by increasing voltage level for power transmission.
- Voltage level of power transmission can be stepped up or down easily with the help of a transformer.

Limitation of AC lines

- At very high voltage, corona loss increases.
- Skin effect is more in AC line. It increases effective resistance.
- Reactive voltage drop is large compared to resistive voltage drop. Therefore, reactive power compensation is required for voltage control.
- Number of conductors and conductor volume is large compared to DC lines.

Advantages of DC lines

- Conductor number and conductor volume are less than AC line.
- Corona loss is very small compared to AC lines.
- Skin effect is less in DC line.
- It does not have any reactive voltage drop. Therefore, voltage stability or regulation problem is less in DC line.
- Single-wire, two-wire and three-wire transmission is possible with or without earth return.

Limitation of DC lines

- Stepping up and down of voltage is more difficult than AC lines.
- Additional rectifier and inverter units are required.
- Converters introduce lots of harmonics.

2.14.2 Overhead lines versus underground lines

Major advantages and limitations of overhead lines and underground lines are as follows:

Advantage of overhead lines

- In overhead lines, air acts as natural insulator between conductors.
- Voltage levels can be made very high.
- Natural air cooling is available. Power transmitting capacity can be increased.
- Visual inspection is possible.
- Cost is less compared to underground lines.

Limitation of overhead lines

- Overhead lines require to maintain minimum gap between ground level and conductor.
- In densely populated area, it becomes difficult to install overhead lines.
- Overhead lines are directly exposed to lightning and other environmental hazards.
- Sometimes it becomes difficult to pass overhead lines over high road or express way, river traction lines, buildings, etc.
- Ice effect and wind effect are present on transmission system.

Advantage of underground lines

- Underground lines can be installed in densely populated area.
- It becomes easier to pass underground lines under high road, river traction lines, buildings, etc.
- Short-circuit probability due to external causes reduces in line.

- It is not directly exposed to lightning.
- There is no ice effect or wind effect.

Limitation of underground lines

- Cost of underground line is high as compared to overhead lines of same voltage level.
- Power transmission at very high voltage is not possible
- Underground lines require insulation over the conductor, which increases cost.
- Sometimes it becomes difficult to get suitable underground path in densely populated city area which may have frequent water lines, communication lines, foundation of civil construction, etc.

2.15 Efficiency

Ratio of output power to input power is known as efficiency.

$$\text{Efficiency } (\eta) = \frac{\text{output power}}{\text{input power}} \tag{2.143}$$

$$\text{Percentage of efficiency } (\eta\%) = \frac{\text{output power}}{\text{input power}} \times 100\% \tag{2.144}$$

Input of a system is the summation of output and loss of the system. So,

$$\begin{aligned} \text{Efficiency } (\eta) &= \frac{\text{output power}}{\text{input power}} \\ &= \frac{\text{input power} - \text{loss}}{\text{input power}} \\ &= 1 - \frac{\text{loss}}{\text{input power}} \end{aligned} \tag{2.145}$$

2.16 Regulation

Regulation of a transmission line is defined as follows:

$$\text{Regulation} = \frac{\text{receiving end voltage at no-load} - \text{receiving end voltage at full-load}}{\text{receiving end voltage at full-load}} \tag{2.146}$$

Percentage regulation of a transmission line is defined as follows:

$$\text{Percentage regulation} = \frac{\text{receiving end voltage at no-load} - \text{receiving end voltage at full-load}}{\text{receiving end voltage at full-load}} 100\%$$

To maintain the voltage profile of power system, it is utmost important to control the voltage at different location. Some may be highlighted as follows:

- Performance of lighting loads: Performance of lighting loads depends on the voltage fed to them. If voltage differs largely from their rated voltage, their performance falls down from the level of satisfaction. Also longevity of the lighting loads depends on the supply voltage.
- Motor loads: For satisfactory operation of motors connected with power system, rated voltage is to be supplied. Change of supply voltage causes shifting of operating point, torque produced, current drawn by the motors, etc.
- Voltage stability: Voltage stability is an important aspect of reliable operation of a power system. Voltage control is thus very important to maintain voltage stability; otherwise, change of load may cause instability.
- Power delivery capacity: Power delivery capacity of transmission system depends on the voltage profile of buses. It makes voltage control important to obtain optimum performance of the system.
- Load frequency control: For load frequency control, bus voltages are to be maintained within range. Therefore, along with frequency control, voltage control is to be performed simultaneously.

2.17 Major sinks of reactive power

Voltage drop and bus voltages depend on the absorption of the reactive power by different parts of the power system. Major sinks of the reactive power may be mentioned as follows:

- Reactance of the generators: Reactance of the generator acts as absorber of reactive power.
- Line inductance: Voltage drop occurs across the inductive reactance of transmission line, and they act as sink of reactive power.
- Motor loads: Motor (induction motor) loads are major sinks of reactive power.
- Lighting loads: Lighting loads act as sink of reactive power.
- Transformer: Transformer absorbs reactive power.
- Fault current: During short circuit, the system draws excessive amount of reactive power.

2.18 Major sources of reactive power

Generation of reactive power may take place in all parts of the power system as follows:

- Excitation system of generation: Exciters of the generators generate reactive power and act as sources of reactive power.
- Line capacitance: Line capacitance acts as a source of reactive power. In LTL and in some MTL, amount of capacitance is high, and therefore, they act as a major source of reactive power in the system.

- Capacitor bank: Capacitor bank connected with the system acts as a source of reactive power to the system.
- Series capacitor: Capacitor connected in series with the line act as a source of reactive power.
- Synchronous condenser: Synchronous motor running at over-excitation behaves like capacitor and act as a source of reactive power.

2.19 Voltage control centres

Voltage control can be performed in various ways at different parts of the power system as follows:

- Generating stations
- Substations
- Load distribution centres
- Application areas
- Feeders

2.20 Major voltage control techniques or equipment

Major voltage control techniques or equipment may be listed as follows:

- Excitation system
- Tap changing transformer
- Synchronous condenser
- Boost up transformer
- Shunt capacitor
- Series capacitor

2.21 Excitation system at generating station

To generate electrical power, any generator needs DC excitation in its field circuit. For this purpose, there must be source of DC excitation. At present, there are various types of excitation system available in generating station. In some special case, permanent magnet is used to provide magnetic flux in the generator. The other major excitation system can be categorized as follows:

1. Main exciter
2. Main exciter–pilot exciter
3. Rectifier as exciter
4. AC exciter with rectifier
5. AC exciter–pilot exciter with rectifier

Generator terminal voltage is decided by the excitation. An open-loop model of the simple excitation system is shown in Figure 2.21.

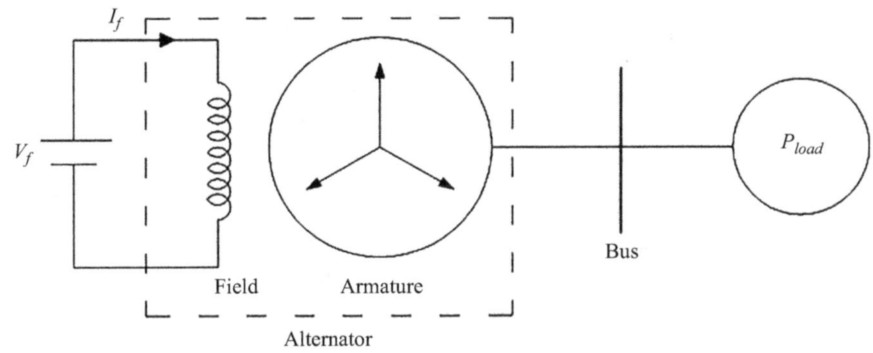

Figure 2.21 Open-loop model of simple excitation system

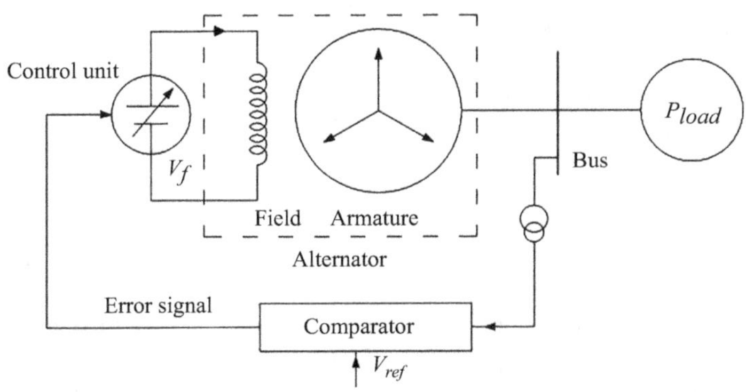

Figure 2.22 Closed-loop model of excitation system

Variation in load changes the terminal voltage in open-loop system. A closed-loop model of excitation system is shown in Figure 2.22.

In a closed-loop model, terminal voltage is stepped down through potential transformer, and here from, error signal is generated. Reference voltage signal (V_{ref}) is used to generate error signal. In the control unit, control signal is generated. The control unit changes the excitation voltage, which regulates the voltage level of generation.

2.21.1 Main exciter

A main exciter model is shown in Figure 2.23. A DC shunt generator is used as a main exciter. Terminals of the DC shunt generator are connected to the terminals of field circuit of the AC generator through a slip ring.

Error signal is generated from the generator bus voltage which is connected with the terminals of the generator. Control unit controls the DC field current of the

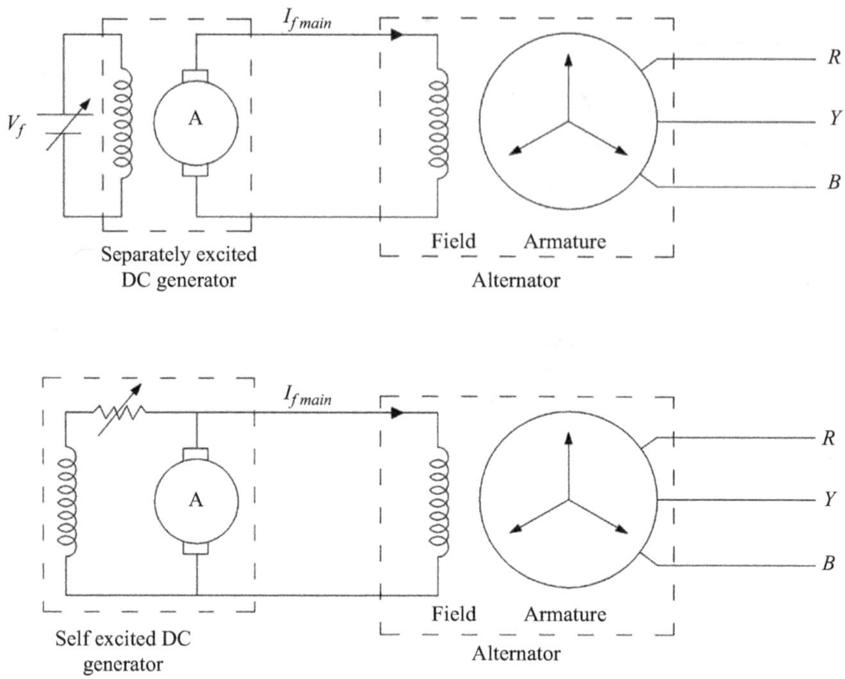

Figure 2.23 Main exciter model

shunt generator. Sometimes, an auxiliary control unit is provided to control the generated voltage of the shunt generator. This auxiliary control unit takes the shunt generator's output as feedback signal. Generated voltage of the shunt generator is fed to the field circuit of the AC generator as excitation.

2.21.2 Main exciter–pilot exciter

A main exciter–pilot exciter model is shown in Figure 2.24. Separately excited a DC generator is used as a main exciter. Terminals of the DC shunt generator are connected to the terminals of field circuit of the AC generator through a slip ring. Excitation to the field circuit of the main exciter is given from another DC shunt generator which is called pilot exciter. Use of the pilot exciter provides better control of excitation. Also, if any one of the two exciters fails, the remaining one can run the system temporarily for short duration.

Error signal is generated from the bus voltage which is connected with the terminals of a generator. The control unit controls the DC field current of the pilot exciter. Sometimes, an auxiliary control unit is provided to control the generated voltage of the pilot exciter. This auxiliary control unit takes the shunt generator's output as feedback signal. Generated voltage of the pilot exciter, i.e. a DC shunt generator, is fed to the field circuit of main exciter, output of which is fed to field circuit of an AC generator as excitation.

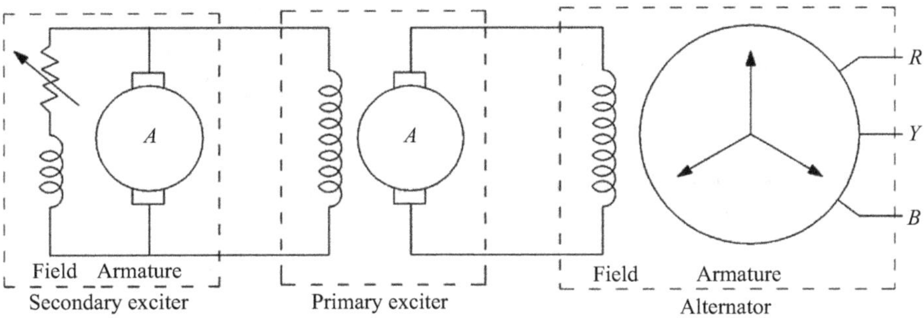

Figure 2.24 Main exciter–pilot exciter model

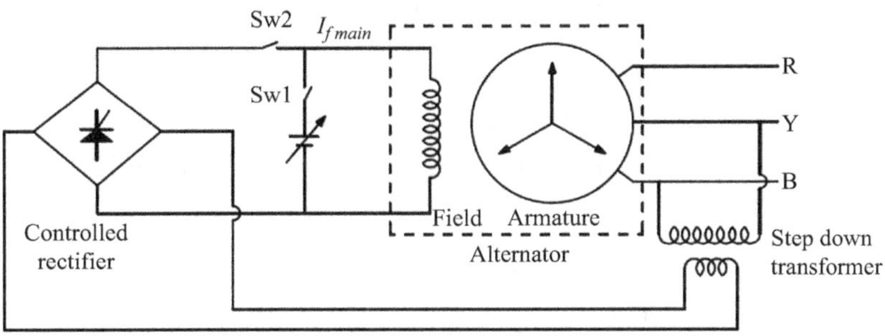

Figure 2.25 Rectifier model of exciter

2.21.3 Rectifier as an exciter

A rectifier model of an exciter is shown in Figure 2.25. A rectifier unit is used as a main exciter. The output terminals of the rectifier are connected to the terminals of the field circuit of the generator through a slip ring.

Error signal is generated from the generator bus voltage which is connected with the terminals of the generator. The control unit controls the rectified output. The AC input of the rectifier is taken from the output of the main AC generator.

2.21.4 AC exciter with a rectifier

An AC exciter with a rectifier model of an exciter is shown in Figure 2.26. The output terminals of the rectifier are connected to the terminals of the field circuit of the generator through a slip ring. The AC input of the rectifier is taken from the output of another AC generator.

Error signal is generated from the generator bus voltage which is connected with the terminals of a generator. The control unit controls the field of the AC generator which supplies AC input to the rectifier. Rectified output is fed to the excitation of the main generator as excitation.

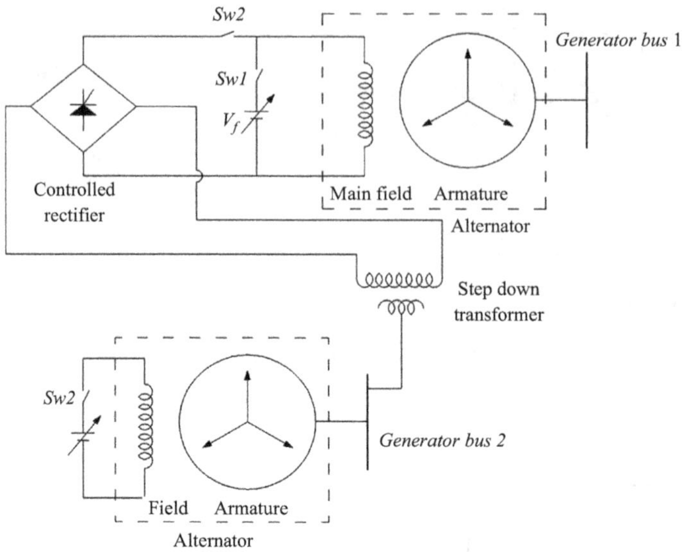

Figure 2.26 AC exciter with rectifier model of exciter

2.21.5 AC exciter–pilot exciter with rectifier

An AC exciter–pilot exciter with a rectifier model of an exciter is shown in Figure 2.27. The output terminals of the rectifier are connected to the terminals of the field circuit of a separately excited DC generator. The AC input of the rectifier is taken from the output of main AC generator. The excitation of the main AC generator is provided by the separately excited DC generator.

Error signal is generated from the generator bus voltage which is connected with the terminals of a generator. The control unit controls the field of a pilot exciter which gives excitation to the AC exciter which supplies AC input to the rectifier. Rectified output is fed to the excitation of the main generator as excitation.

2.22 Tap changing transformer

2.22.1 Position of high-voltage winding

Transformer windings are divided into two windings – primary and secondary. These two windings are magnetically linked through a magnetic core material. Two types of transformer's construction are available: core type and shell type. Position of HV and LV windings of two types of transformers are shown in Figures 2.28 and 2.29, respectively. In both the cases, HV winding is placed in the outer part which has higher volume and surface area with more insulating materials.

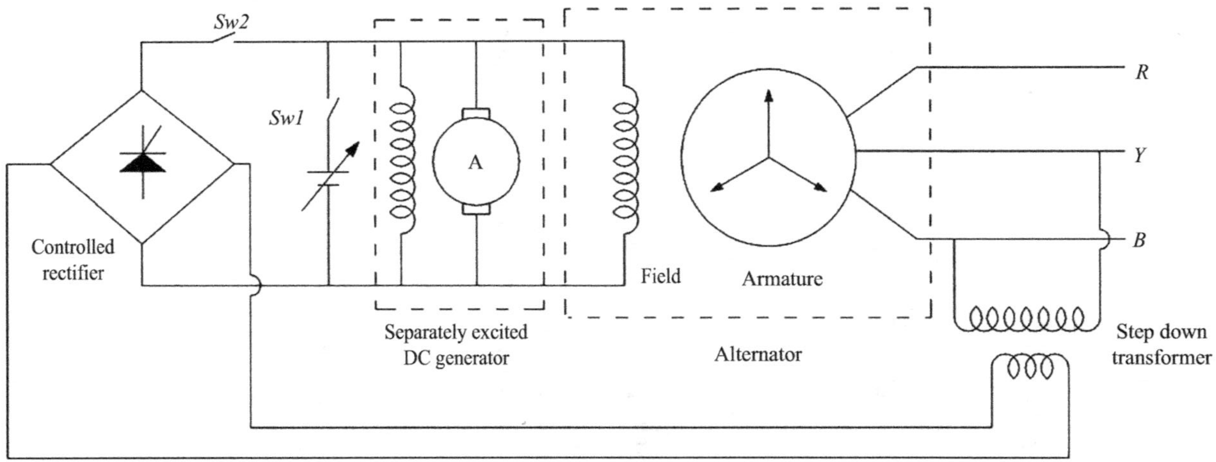

Figure 2.27   AC exciter–pilot exciter with rectifier

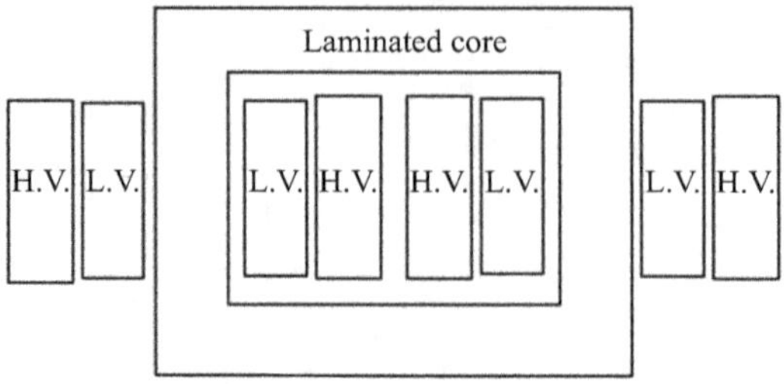

Figure 2.28 Core-type transformer

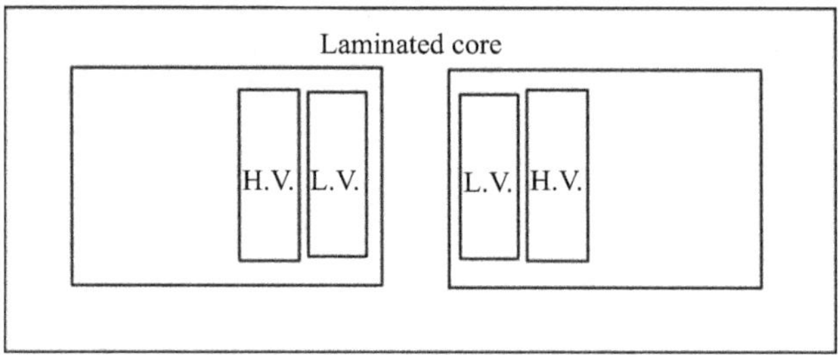

Figure 2.29 Shell-type transformer

2.22.2 Transformer operation

Considering different design factors as unity, voltage across primary and secondary may be written as follows:

$$V_P = 4.44 f \varphi_m N_P \tag{2.147}$$

$$V_S = 4.44 f \varphi_m N_S \tag{2.148}$$

where N_P is the number of turns in primary winding, N_S is the number of turns in secondary winding.

Therefore,

$$V_S \propto V_P \tag{2.149}$$

and

$$\frac{V_S}{V_P} = \frac{N_S}{N_P}$$

or,

$$\frac{V_S}{V_P} = \frac{I_P}{I_S} = \frac{N_S}{N_P} \tag{2.150}$$

2.22.2.1 Turns ratio

The turns ratio refers to the ratio of the number of turns in secondary winding to the number of turns in primary winding which is approximately equal to secondary-to-primary voltage ratio.

$$\text{Turns ratio } (N) = \frac{N_S}{N_P} \approx \frac{V_S}{V_P} \tag{2.151}$$

2.22.2.2 Equivalent circuit of a transformer at no-load

Equivalent circuit at no-load has been shown in Figure 2.30 and the corresponding phasor at no-load has been shown in Figure 2.31, where R_P is the resistances of primary winding, R_S is the resistances of secondary winding, X_P is the leakage reactance of primary winding, X_S is the leakage reactance of secondary winding, I_0 is the exciting or no-load current, I_m is the magnetizing component of exciting or no-load current and I_w is the loss component of exciting or no-load current, R_m is the core-loss resistance through which I_w flows, X_m is the magnetizing reactance through which I_m flows, E_P is the back emf in the primary side and E_S is the induced emf in the secondary side.

Magnetizing current is the minimum current required to build up flux distribution in the core of a transformer.

2.22.2.3 Equivalent circuit of a transformer at load

An equivalent circuit of a transformer and the corresponding phasor diagram at load are shown in Figures 2.32 and 2.33, respectively. Here, load impedance is Z_L.

Voltage regulation of a transformer may be defined as follows:

$$\text{Voltage regulation} = \frac{(V_{S\text{-no-load}} - V_{S\text{-full-load}})}{V_{S\text{-full-load}}} \tag{2.152}$$

where,

$V_{S\text{-no-load}}$ = voltage across secondary at no-load
$V_{S\text{-full-load}}$ = voltage across secondary at full-load

$$\text{Percentage voltage regulation} = \frac{(V_{S\text{-no-load}} - V_{S\text{-full-load}})}{V_{S\text{-full-load}}} \times 100\% \tag{2.153}$$

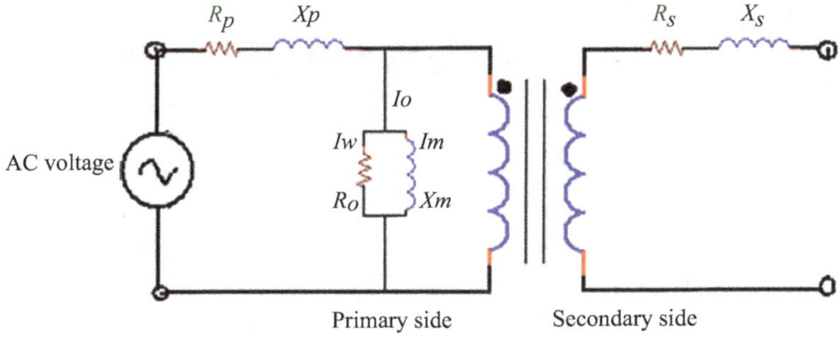

Figure 2.30 Equivalent circuit of a transformer at no-load

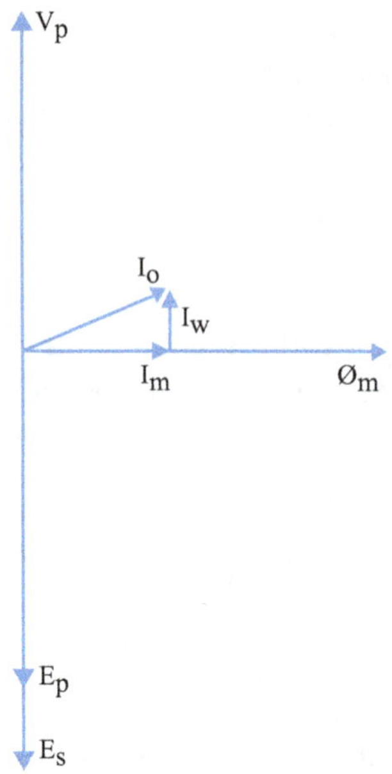

Figure 2.31 Phasor diagram of transformer at no-load

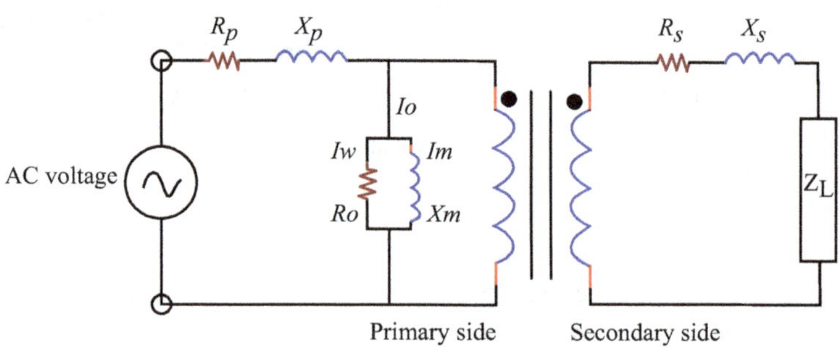

Figure 2.32 Equivalent circuit of a transformer at load

Let, I_P be the current through primary winding and I_S be the current through secondary winding.

Therefore, volt–ampere relation between primary and secondary can be written as follows:

$$V_P I_P \approx V_S I_S$$

Also,

$$\text{Turns ratio } (N) = \frac{N_S}{N_P} \approx \frac{V_S}{V_P} \tag{2.154}$$

Thus, voltage across secondary terminals depends on the load and also on the turns ratio. Therefore, by varying the turns ratio, terminal voltage can be controlled.

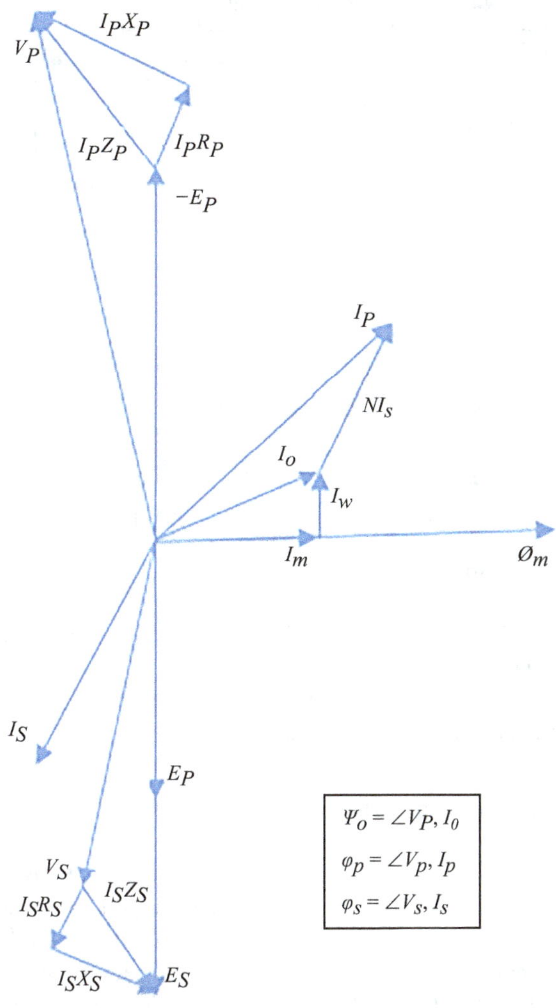

Figure 2.33 Phasor diagram of transformer at load

To change the turns ratio, tapping is provided. In order to simplify calculation, it is theoretically possible to transfer voltage, current and impedance of one winding to the other and combine them into single values for each quantity. Thus, we can work into one winding only which is more convenient. These parameters can be referred to primary or secondary side. Equivalent circuit of a transformer with load in referred to primary has been shown in Figure 2.34.

2.22.3 Off- and on-load tap changing

Tap changing can be done in off- or in on-load conditions. On-load tap changing is associated with arcing; care is to be taken against arcing, and some method can be adopted to minimize arcing. However, to avoid this problem, off-load tap changing method is followed.

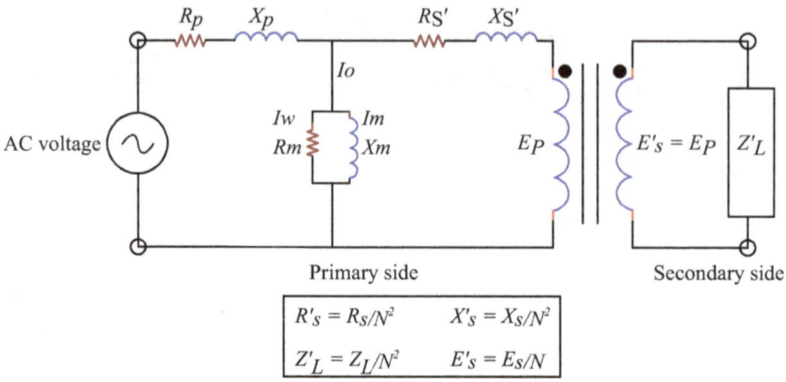

Figure 2.34 Equivalent circuit of a transformer at load in referred to primary

2.22.4 Location of tapping

Tapping is placed on the HV side. Major reasons for this are as follows:

- Windings of HV side have a higher number of turns. So, better resolution can be achieved if tapping is provided on the HV side.
- Current in HV side is lower than current of the LV side, which results in less arc related hazards.
- HV winding is placed on the outer part of the winding. So, it is easier to connect at the outer part of the winding.

2.23 Synchronous machine

A synchronous machine that converts the mechanical power from a prime mover into an AC electrical power at a particular voltage and frequency is known as synchronous generator or alternator. An alternator has field winding on the rotor and three-phase armature winding on the stator. The three-phase armature windings are place at 120° apart from each other on the stator slots. The rotor is coupled with turbine set. Mechanical energy is fed to the rotor through turbine. When the rotor rotates, DC supply is fed to the field winding. This is shown in Figure 2.35.

Field current produces magnetic flux in the air gap and links the armature winding. As the rotor rotates, flux linkage occurs on the armature winding. Due to the relative motion of the flux and armature winding, electromotive force (EMF) is induced in the winding. As they are 120° apart from each other, the EMFs induced in the three windings of the armature are at 0°, 120° and 240° in respect of phase. The shapes of the generated EMFs are shown in Figure 2.36. Phasors of generated voltages are shown in Figure 2.37.

The equivalent circuit of an alternator is shown in Figure 2.38. Here, field winding is fed by DC supply and the three-phase armature winding is shown by single-phase winding having resistance R_S and reactance X_s, and they are in series. The generated EMF per phase is shown connecting an AC voltage source in series with R_S and X_s.

Synchronous reactance: It is the reactance contributed by the inductance property of armature winding of the alternator.

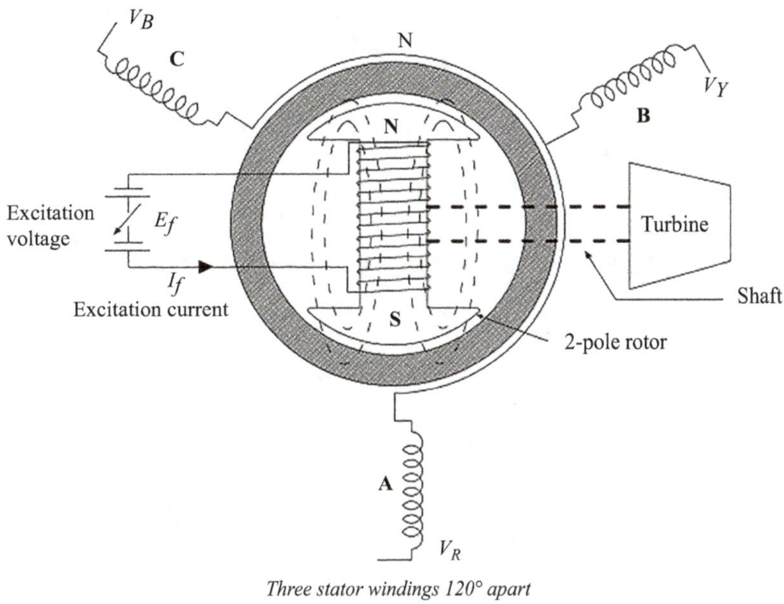

Figure 2.35 Alternator with DC excitation and mechanically coupled with a prime mover and turbine arrangement

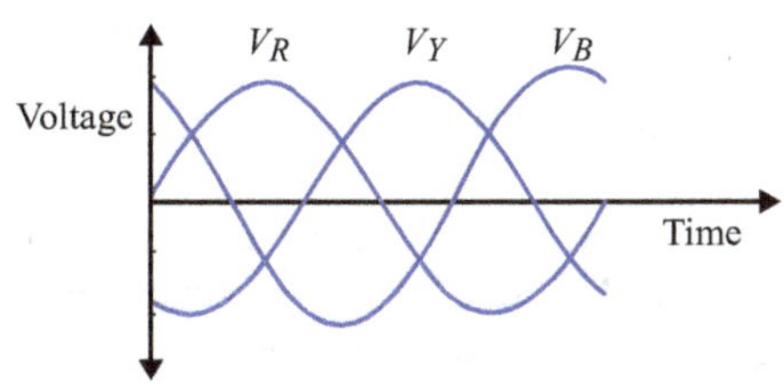

Figure 2.36 Three-phase generated voltages

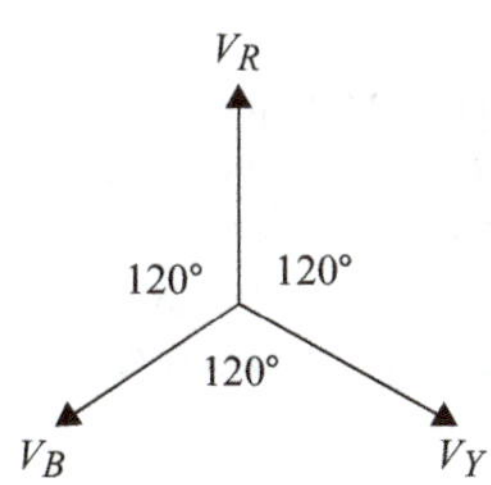

Figure 2.37 Phasors of generated voltages

Synchronous impedance: It is the impedance constituted by armature resistance and synchronous reactance.

$$Z_s = R_S + jX_s \tag{2.155}$$

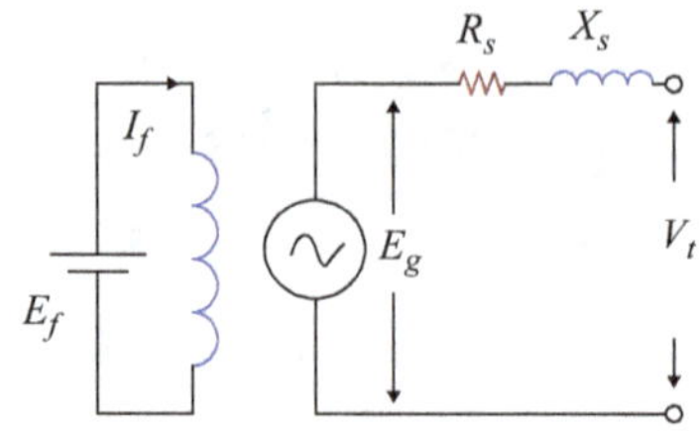

Figure 2.38 Equivalent circuit of an alternator

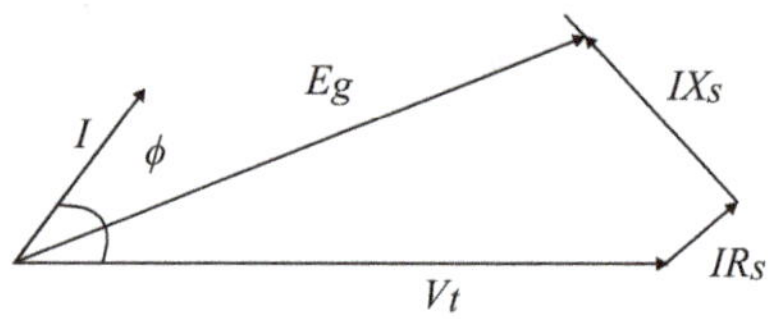

Figure 2.39 Phasor at capacitive load

Phasors with leading current is shown in Figure 2.39. With leading current, terminal voltage (V_t) exceeds generated voltage (E_g). But in reality, power system load is inductive, and at load terminals, voltage remains less than generated voltage. Power engineers work to regulate the voltage profile considering the phasor relation of terminal voltages with the nature of loads.

2.24 *I–V* Characteristics without voltage control

With the increase of lagging load, voltage drop increases. It results in a reduction of terminal voltage as shown in Figure 2.40. Similarly, with the increase of leading load, voltage drops become additive with sending-end voltage, and it results in an increase of terminal voltage.

2.25 *I–V* Characteristics with ideal voltage control

In a power transmission system having adequate provision of control devices for voltage/reactive power/power factor control, during lagging load, reactive power is supplied, and during leading load, reactive power is absorbed. This keeps the voltage within a range. For transmission system with ideal reactive power control system, voltage remains constant and does not change with load irrespective of its nature as shown in Figure 2.41.

2.26 *P–V* Characteristics

Consider a transmission line as shown in Figure 2.42. It is connected with a source which provides sending-end voltage V_S. Receiving-end voltage is V_R, and the line is

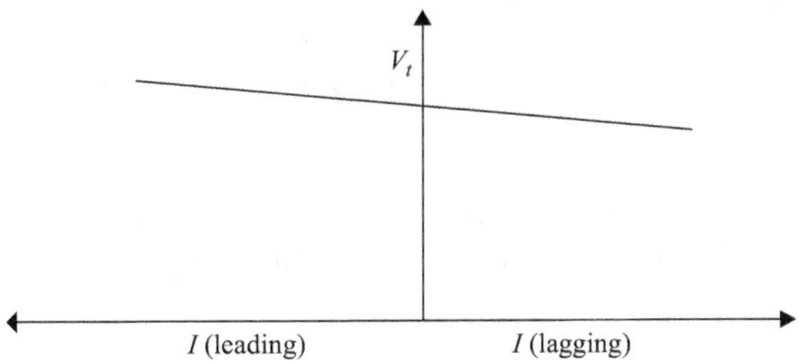

Figure 2.40 I–V Characteristics without voltage control

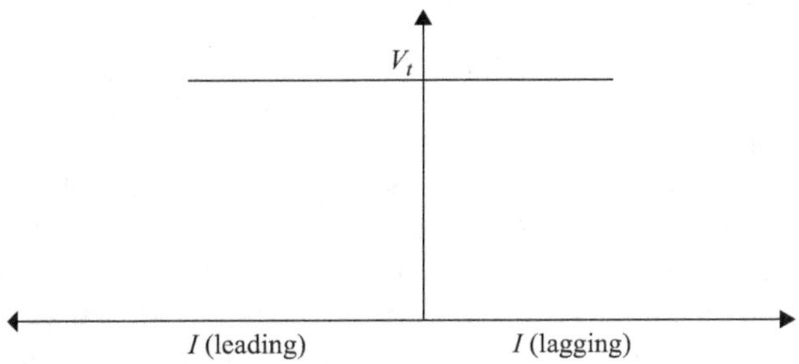

Figure 2.41 I–V Characteristics with ideal voltage controller

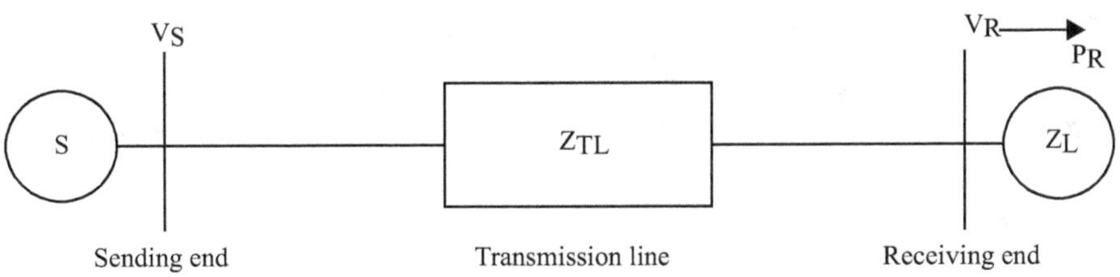

Figure 2.42 Transmission line connected with load and transmitting active power

transmitting active power P_R. Let $P_{\max}$ be the maximum power that can be transmitted at unity power factor at stable condition. Line impedance and load impedance are as follows:

Transmission line impedance $= Z_{TL} = R + jX_{TL}$

Load impedance $= Z_L = R + jX_L$

Now, theoretically following four conditions may occur:

1. No-load
2. Loading with unity power factor (UPF)
3. Loading with lagging power factor
4. Loading with leading power factor

2.26.1 No-load

At no-load, there is no active power transmission, i.e.

$$P_R = 0$$

Therefore, there will not be any reactive drop (neglecting charging current), and receiving-end voltage (V_R) will become equal to sending voltage (V_S), i.e.

$$V_S = V_R$$
$$\therefore \frac{V_R}{V_S} = 1$$

2.26.2 Loading with unity power factor

With the increase of load at UPF, active power transmission will increase. This will make more voltage drop across line reactance and will reduce receiving-end voltage. Thus, with the increase of P_R/P_{max} ratio, V_R/V_S ratio will decrease as shown in Figure 2.43. This will continue until active power transmitted through the line becomes equal to P_{max}. After that, stability will be lost causing reduction of voltage, making the system incapable of generating and transmitting more active power. As a result, both receiving-end voltage (V_R) and amount of transmitted power will reduce simultaneously to minimum or zero value, making the system more unstable. That is why the point corresponding to is called critical point, and this point defines the maximum power that can be transmitted at stable condition beyond which the system lost stability.

2.26.3 Loading with lagging power factor

For loading with lagging power factor, maximum power is less than that of UPF. Therefore, with the increase of active power transmission, receiving-end voltage (V_R)

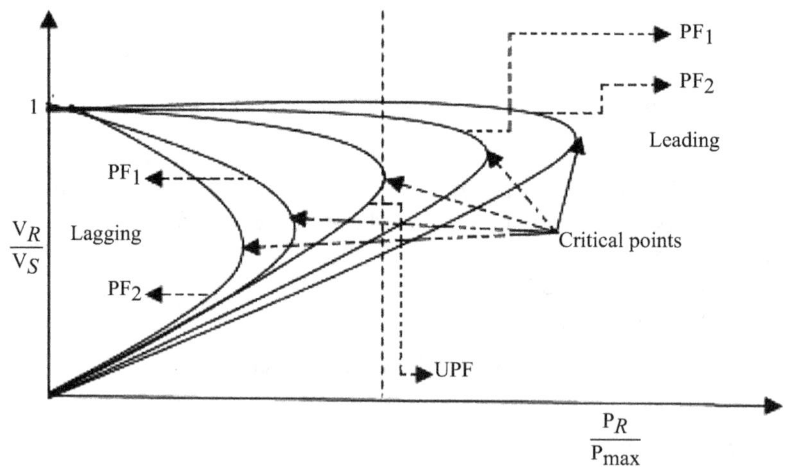

Figure 2.43 P–V characteristics

decreases more, and critical point is reached before that of UPF. If the system is more inductive, power factor will be more lagging reaching critical point early.

2.26.4 Loading with leading power factor

For loading with leading power factor, maximum power is more than that of UPF. Therefore, with the increase of active power transmission, receiving-end voltage (V_R) decreases less and critical point is reached after that of unit power factor.

2.27 Voltage, power and impedance

With the increase of load, load impedance or load reactance will reduce. This will increase the ratio Z_{TL}/Z_L. Therefore, with the increase of active power transmission, the ratio Z_{TL}/Z_L will increase, and receiving-end voltage (V_R) will decrease as shown in Figure 2.44. When power transmission becomes maximum, it touches critical point as shown in Figure 2.45 and Figure 2.46. After that, with the increase of the ratio Z_{TL}/Z_L, system becomes unstable and system voltage cannot be restored. Corresponding impedance versus current characteristic has been shown in Figure 2.47. Reactive power with respect to voltage for different active power has been shown in Figure 2.48.

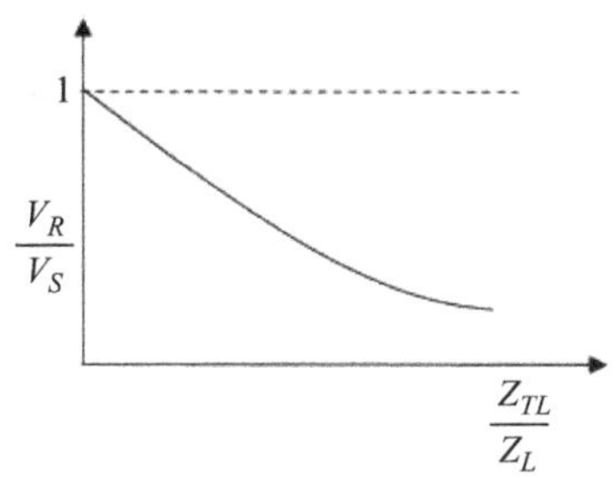

Figure 2.44 Impedance versus voltage characteristics

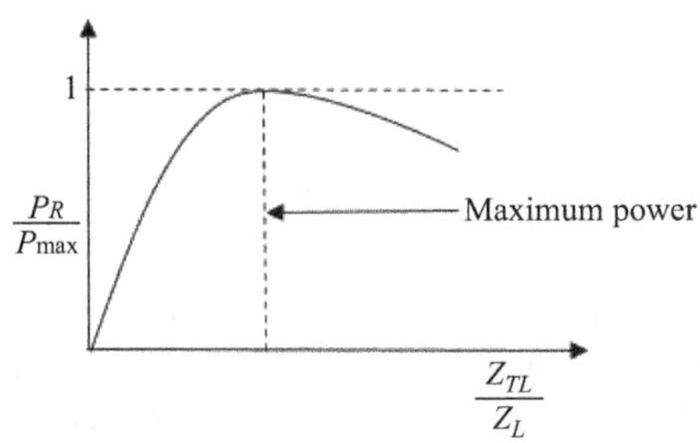

Figure 2.45 Impedance versus active power characteristics

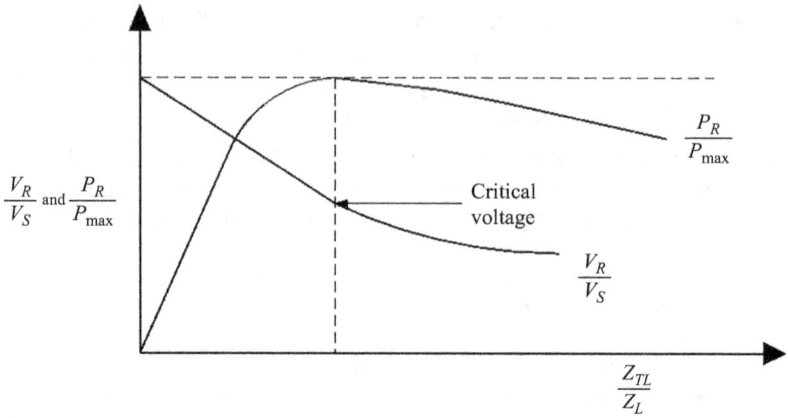

Figure 2.46 Impedance versus voltage and active power characteristics

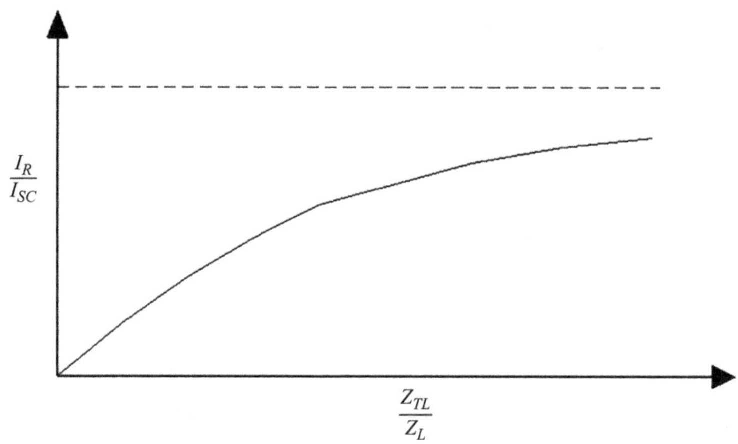

Figure 2.47 Impedance versus current characteristics

2.28 Synchronous condenser

A synchronous motor at over-excitation draws leading current. For this feature, the synchronous motor is used at over-excitation to act as a source of reactive power to power system network. Synchronous motor drawing leading current is known as synchronous condenser. Field current versus armature current of synchronous

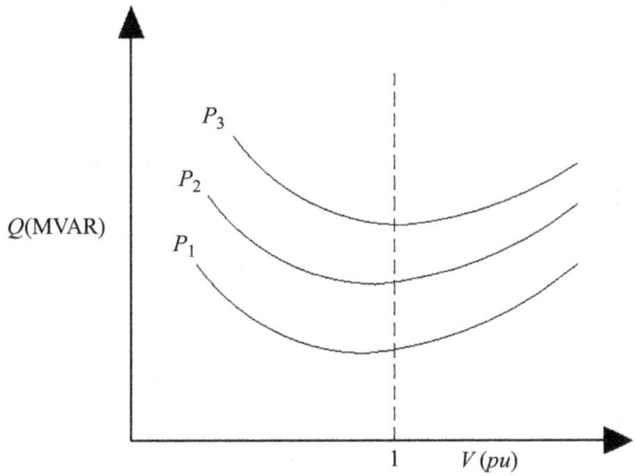

Figure 2.48 Q–V Characteristics for different active powers

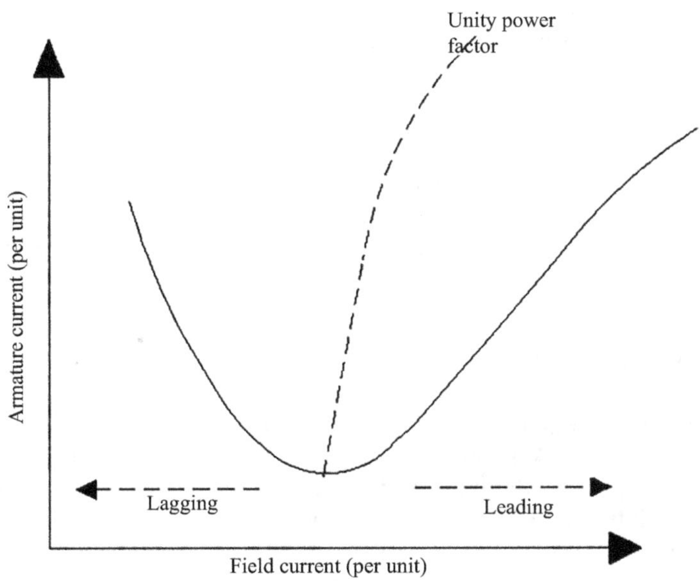

Figure 2.49 V-Curve of synchronous motor

motor looks like 'V' shaped as shown in Figure 2.49; the characteristics are known as the V-curve. It shows that armature current reduced with the increase of field current at a lagging power factor and armature current increases with the increase of field current at a leading power factor.

2.29 Voltage collapse

Voltage collapse refers to lowering of bus voltage level below acceptable range leading the power system into voltage instability as a result of single or series of events. It may occur within several minutes and may exist for few seconds to a large period of time. The increase of reactive power demand is the main reason for voltage collapse. There are many reasons for which voltage collapse may occur. Major causes for voltage collapse are mentioned as follows:

- Increase of reactive power demand for increase of load
- Fault
- Sudden loss of one EHV line from the system
- Failure of tap changing operation of transformer
- Loss of excitation in alternator operation
- Failure in voltage control mechanism
- Failure in reactive power compensator
- Failure of coordination between voltage control unit protective devices

Besides the previous reasons, there are various factors on which voltage collapse in a power system depends. Some of them are mentioned as follows:

- Type of bus
- Bus location
- Distance between bus and load
- Distance between bus and source of reactive power
- Distance between bus and location of voltage or reactive power controllers
- Line parameters

2.30 Voltage stability

Power system network is facing ever-increasing power demand. Thus, it has become a difficult task to make the operation stable and to improve power transmission capacity. Many alternate solutions have been introduced to cope with this like introduction of parallel circuit, extra high voltage (EHV) line and high voltage DC (HVDC) transmission. However, with the development of power electronic switching devices, better solution has been introduced which compensates reactive power demand as required maintaining the voltage profiles and angle within desired ranges increasing power transmission capacity. This system is named flexible AC transmission system (FACTS) which offers improved power transmission capacity and maintains voltage profile within range, meeting reactive power demand. Different types of control devices have been introduced which will be discussed in the following sections.

2.31 Factors of power transmission capacity

Power engineers aim to maximize power transmission capacity of the transmission line keeping the operation stable. However, the task is difficult and various factors are there to restrain the system from maximizing power transmission capacity beyond any limit.

With the increase of power transmission, mechanical dimension and weight of the conductor increase, which also increases construction cost to carry them.

With the very increase of transmission voltage, corona loss increases. Also HV requires more line insulators. With the increase of transmission voltage, clearance from the ground level and inter-distance between any two conductors increase; sometimes these may not be feasible and cost effective for transmission line covering short or medium distance and located in dense populated areas.

To overcome these limitations, three options have been introduced: parallel power transmission, high voltage DC power transmission and flexible AC power transmission.

2.31.1 Parallel power transmission

In parallel transmission, more power is transmitted in two or more parallel lines. As a result, mechanical dimension and weight of the conductor do not increase too high. However, parallel power transmission has also limit beyond which power is not transmitted.

2.31.2 High-voltage DC (HVDC) power transmission

Loss of power transmission in high or extra HVAC mode can be reduced by converting it into DC and carrying power in HV. In HVDC transmission, corona loss reduces drastically which improves power transmission capacity. However, HVDC transmission is not a good solution for short and medium transmission lines.

2.31.3 Flexible AC power transmission

In flexible AC power transmission reactive power, voltage angle and power factor are controlled in various ways keeping the system stable and improving power transmission capacity. Moreover, necessary control may be provided from generation side as well as from transmission and distribution sections of the power system. With the progress of power electronics and computer technology, different FACTS controllers have been introduced to provide better solution, making this option more popular.

2.32 Flexible AC transmission system

AC transmission system having extended capacity of power transmission and extended ability to operate at stable condition during gradual small change as well as sudden high change of load is known as FACTS. FACTS provides better control of one or more AC transmission system parameters to enhance controllability and increase power transfer capability. This is achieved by delivering or absorbing reactive power through power-electronic-based switching devices. Use of semiconductor-based switching devices increases controllability of the overall transmission system. Thus, the AC transmission system can cope with the different power transmission condition and become more flexible with respect to all matters related to power transmission. Switching circuits, power sources and other related components used in flexible AC transmission system are known as FACTS devices.

Switching circuits, voltage and current controllers, inverters, power sources, etc. are essential parts of FACTS devices. Two types of inverter are used as follows:

1. Voltage source inverter (VSI)
2. Current source inverter (CSI)

In VSI, direct voltage is inverted and fed to AC system. In CSI, direct current is inverted and fed to AC system. In FACTS controllers, power sources are not mandatory. However, when they are used, they are either voltage source or current source. These sources may be fixed or may be rectified source. In FACTS operation, handling of change of direction of power is very important. For FACTS having VSI, change of direction of power is done by reversal of phase current, whereas, for FACTS having CSI, change of direction of power is done by reversal of phase of current.

2.32.1 Power and reactance

Let, δ is the load angle.

$$P = \left(\frac{EV}{X}\right)\sin(\delta) \tag{2.156}$$

$$V_S = V\cos\left(\frac{\delta}{2}\right) + jV\sin\left(\frac{\delta}{2}\right) \tag{2.157}$$

$$V_r = V\cos\left(\frac{\delta}{2}\right) - jV\sin\left(\frac{\delta}{2}\right) \tag{2.158}$$

$$I = \frac{V_S - V_r}{jX} = \frac{2V\sin\left(\frac{\delta}{2}\right)}{X} \tag{2.159}$$

For no-loss line, active power P will be same at any point of the line:

$$P_S = P_R = P = \left(\frac{EV}{X}\right)\sin(\delta) = \frac{V\cos\left(\frac{\delta}{2}\right)2V\sin\left(\frac{\delta}{2}\right)}{X}\sin(\delta) = \frac{V^2}{X}\sin\delta \tag{2.160}$$

Reactive power at sending end will be opposite of reactive power at receiving end:

$$Q_S = -Q_R = Q = \frac{V\cos\left(\frac{\delta}{2}\right)2V\sin\left(\frac{\delta}{2}\right)}{X} = \frac{V^2}{X}(1-\cos\delta) \tag{2.161}$$

As δ is very small, active power mainly depends on δ, whereas reactive power mainly depends on voltage magnitude.

2.32.2 Main features of FACTS

Main features of FACTS devices can be summarized as follows:

1. FACTS devices consist of switching circuits. These switching circuits are made of semiconductor-based power electronics components which give better controllability.
2. FACTS devices include phase controller. Depending upon phase relation between voltages and currents delivering and absorption of reactive power, direction of flow of power and involvement of active power are decided.
3. FACTS may or may not include power source.
4. FACTS devices include inverter which may be either VSI or CSI.

2.32.3 Merits of FACTS

Three major achievements by the application of FACTS are as follows:

1. Dynamic voltage control: The overvoltage that occurs in long lightly loaded lines and cables is limited by FACTS devices. Decrease of voltage, voltage collapse in heavily loaded or faulty lines are limited by FACTS devices.
2. Power transmission capacities: Without adding new lines, application of FACTS devices increases power transmission capacities.
3. Addition of reactive power: The application of FACTS devices makes it easy to add new renewable energy sources to the existing grid system.

Some specific merits obtained from FACTS may be summarized as follows:

- Improved voltage profile of transmission line buses
- Improved reactive power profile
- Improved phase controllability
- Improved power transmission capacity
- Improved steady-state and transient stability margin
- Improved loadability
- Controlled series voltage drop
- Less reactive power burden
- Improved voltage security
- Less probability of voltage collapse

2.32.4 Classification of FACTS devices

Controller devices used in FACTS can be broadly classified into following categories:

1. Series controller
2. Shunt controller
3. Series–series controller
4. Series–shunt controller

Commonly used series controllers are as follows:

1. Static synchronous series compensator (SSSC)
2. Thyristor-controlled series capacitor (TCSC)
3. Thyristor switched series capacitor (TSSC)

Commonly used shunt controllers are as follows:

1. Static VAR compensator (SVC)
2. Static synchronous compensator (STATCOM)
3. Static synchronous generator (SSG)

Commonly used series–shunt controllers are as follows:

1. Static phase shifter (SPS)
2. Unified power flow controller (UPFC)

2.32.5 Series controller

Connection of series controller is shown in Figure 2.50. Here series controller is connected in series with transmission line in between two buses.

It offers variable series impedance. Thus, line voltage drop can accordingly be varied. When load varies, current changes and drop varies, which can be changed as desired for better flexibility. Depending upon requirement, series controller changes voltage drop and may deliver reactive power to the system or absorb reactive power from the system. When voltage is in quadrature with phase of current, only reactive power is involved in operation in terms of delivering or absorption of reactive power. When voltage is not in quadrature with phase of current, only reactive power is involved in operation in terms of delivering or absorption of reactive power along with some real or active power. Series controller is costly and difficult to implement in compared to shunt controller in live system.

Equivalent circuit of a capacitive series controller is shown in Figure 2.51.

With the increase of length of transmission line, line inductance increases. It increases inductive drop. Therefore, connection of series capacitance decreases the line voltage drop.

Series controller changes line impedance. Capacitive series reactance decreases net reactance and hence increases transmittable active power. It also increases the amount of reactive power of the system.

$$P = \frac{V^2}{X - X_C} \sin \delta \tag{2.162}$$

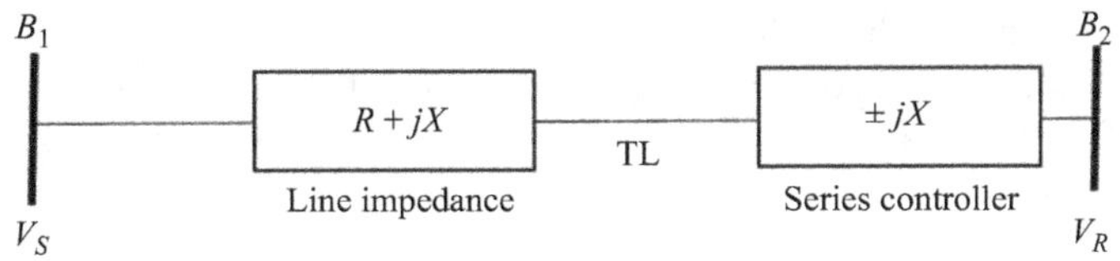

Figure 2.50 Series controller

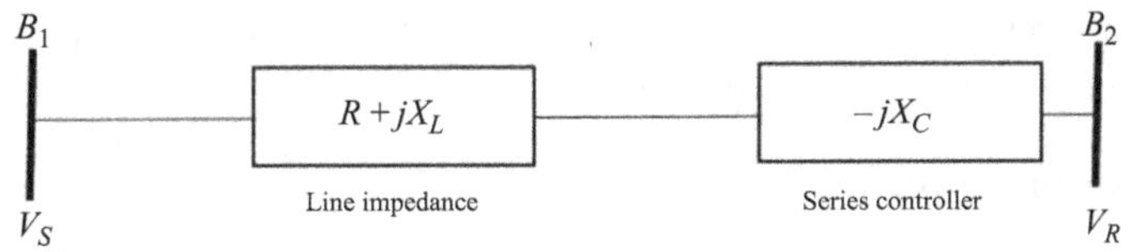

Figure 2.51 Equivalent circuit of a capacitive series controller

$$Q = \frac{V^2}{X - X_C}(1 - \cos\delta) \tag{2.163}$$

Main features of series controller

Main features of series controller are as follows:

1. It offers variable series impedance.
2. It offers variable series voltage drop.
3. It delivers reactive power to the system or absorbs reactive power from the system depending upon the requirement.

Demerits of series controller

When a bus is connected to two more number of buses, for each line, separate series controller is needed; a common series controller cannot be used.

Series controller devices

Main series controller devices are as follows:

- Thyristor-controlled series reactor
- Thyristor-switched series reactor (TSSR)
- Thyristor controlled series capacitor (TCSC)
- Thyristor-switched series capacitor (TSSC)
- Static synchronous series compensator (SSSC)

Static synchronous series compensator

One of the key FACTS devices is SSSC. Sometimes it is known as S^3C. It consists of a voltage-sourced converter (VSC) and a transformer.

A voltage of variable magnitude in quadrature with the line current is injected by SSSC. Hence, it inserts an inductive or capacitive reactance and influences the transmitted electric power. The VSC unit consists of capacitor. Thus, SSSC is categorized into two groups SSSC without storage and SSSC with storage. Often it consists of varistor, fast protective device, bypass switch and damper circuit.

- Varistor is main protective device. It is ZnO type. It limits voltage across capacitor in safe range and the fault current flowing through the line.
- Fast protective device gives backup protection to the series capacitor when varistor fails to absorb sufficiently high fault current.
- Bypass switch is used to bypass or insert the series capacitor as required.
- Damper circuit is used to damp high-frequency discharge current through fast protective device or through switch.

2.32.6 Shunt controller

Connections of different types of shunt controllers are shown in Figure 2.52. Here the shunt controller is connected to a bus in parallel with transmission line.

It offers variable shunt impedance. This shunt controller draws current from the system, and hence depending upon systems requirement, shunt controller may deliver reactive power to the system or absorb reactive power from the system. When voltage is in quadrature with phase of current, only reactive power is involved in operation in terms of delivering or absorption of reactive power. When voltage is not in quadrature with phase of current only reactive power is involved in operation in terms of delivering and absorption of reactive power along with some real or active power. Shunt controller is also very cost effective.

Shunt compensation may be inductive compensation or capacitive compensation. Shunt unit draws current from the system. STL has negligible line capacitance; MTL has small line capacitance; LTL has considerable high line capacitance. Also in power system, load is inductive in nature. When transmission line is loaded, it becomes inductive and as a result receiving-end voltage decreases. Insertion of capacitive shunt reactance draws leading current, increases power factor and hence improves voltage profile.

However, when LTLs and MTLs become no loaded or low loaded, the overall systems become capacitive. As a result, receiving-end voltage becomes greater than sending-end voltage. It is known as Ferranti effect. It is not also desirable with respect to the aspect of voltage stability, security and insulation. Insertion inductive shunt reactance draws lagging current, decreases power factor and hence decreases receiving-end voltage.

Shunt reactor controls power factor and magnitude of voltage. For more transmittable active power, more reactive power is needed. However, in this case, series reactance will remain same.

$$P = \frac{2V^2}{X} \sin \frac{\delta}{2} \tag{2.164}$$

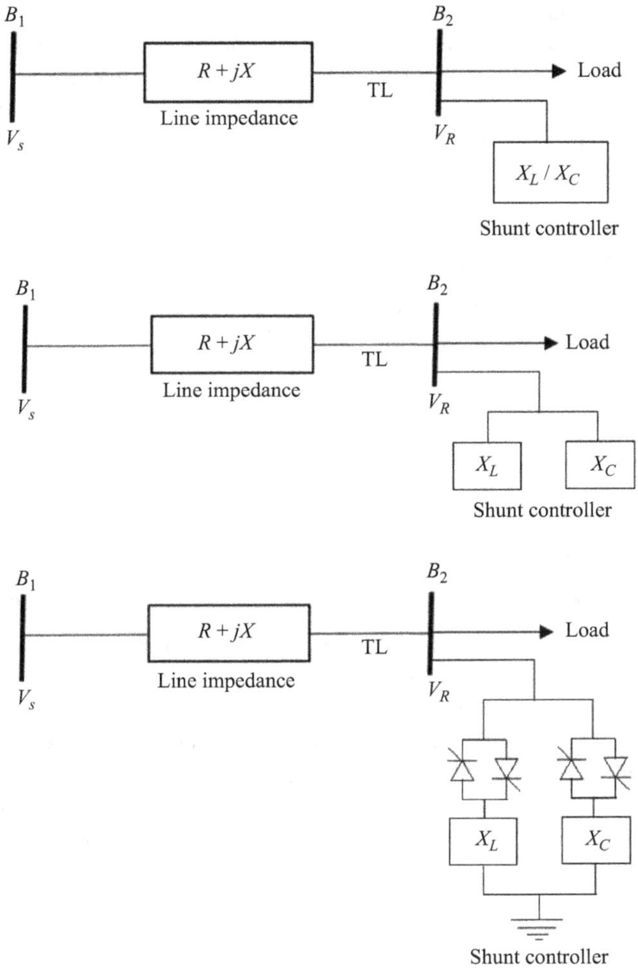

Figure 2.52 Shunt controller

$$Q = \frac{2V^2}{X}\left(1 - \cos\frac{\delta}{2}\right) \tag{2.165}$$

Main features of shunt controller

Main features of shunt controller are as follows:

1. It offers reactive powers variable shunt impedance.
2. It delivers reactive power to the system or absorbs reactive power from the system depending upon the requirement.

Merits of shunt controller

When a bus is connected to two more number of buses, for each line, separate shunt controller is not needed; a common shunt controller is sufficient for VAR control to maintain voltage profile and voltage stability at different load condition.

Common shunt controllers used as FACTS devices are as follows:

1. Static VAR compensator (SVC)
2. Static compensator (STATCOM)
3. Static synchronous generator (SSG)

2.32.6.1 Static VAR compensator

Most common SVCs are as follows:

1. Thyristor-controlled reactor (TCR) and Thyristor controlled reactor / fixed capacitor (TCR/FC)
2. Thyristor-switched reactor
3. Thyristor-switched capacitor (TSC)
4. Mechanically switched capacitor
5. Thyristor controlled reactor / thyristor switched capacitor (TCR/TSC)

2.32.6.2 Thyristor-controlled reactor

Reactor is connected in series with a bidirectional thyristor as shown in Figure 2.53. The thyristor is phase controlled. Equivalent reactance is varied continuously.

Power is controlled by controlling current through the reactor with the help of thyristor switching circuit. On-state duration is controlled by varying the firing angle measured with respect to zero crossing point. In some application, TCR is used with fixed capacitor, i.e. TCR/FC. Use of fixed capacitor helps in reactive power generation. Thus, both reactive power generation and absorption can be

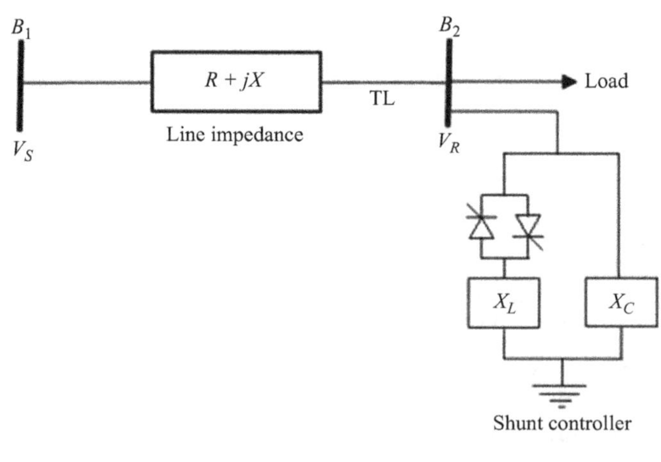

Figure 2.53 Thyristor controlled reactor / fixed capacitor (TCR/FC)

obtained, if required. TCR/FC are used in application of sub-transmission and distribution area.

Main features of TCR/FC are as follows:

1. Control in continuous manner
2. Very little or no transients
3. FC can also be used as filter and helps in harmonic elimination
4. Compact design can be achieved

2.32.6.3 Thyristor-switched reactor

Same as TCR but thyristor is either in zero or full conduction as shown in Figure 2.54. Equivalent reactance is varied in stepwise.

2.32.6.4 Thyristor-switched capacitor

Capacitor bank is connected in series with a bidirectional-thyristor-based switching circuit as shown in Figure 2.55. Thyristor is either in zero or full conduction. Equivalent reactance is varied in stepwise manner. Each branch capacitor is connected with separated switching circuit, and thus, each capacitor can be switched on or off individually. Sometimes switching is done when the voltage across thyristor is zero. It helps in making the system transient free.

Main feature of TSC

Main feature of TSC are as follows:

1. Stepped control is achieved
2. No or little transient
3. Harmonics are present at very small amount
4. Operates with very low loss
5. Flexibility and redundancy are high

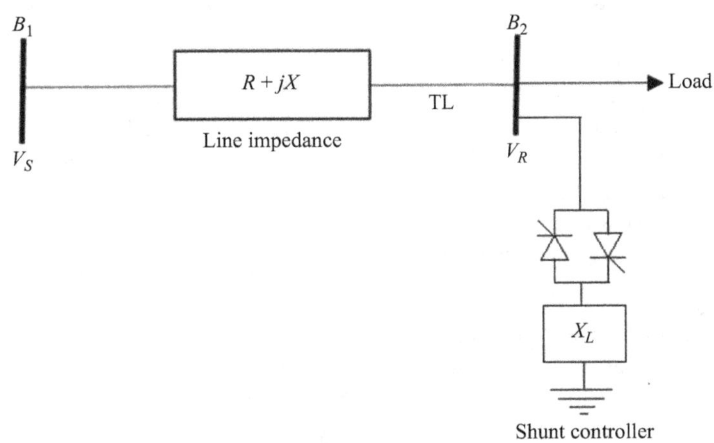

Figure 2.54 Thyristor-switched reactor (TSR)

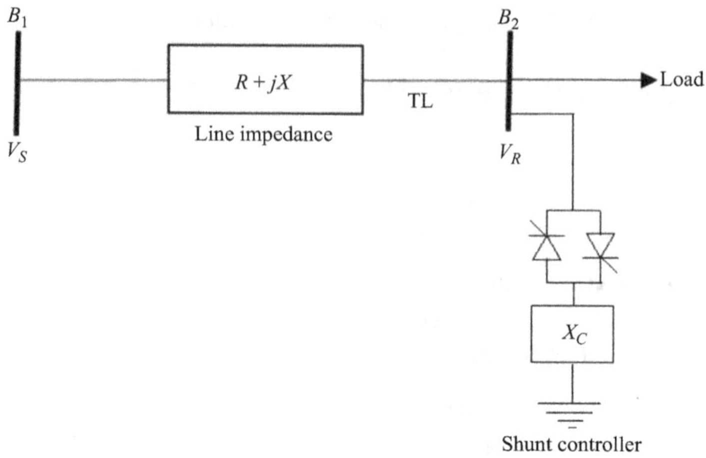

Figure 2.55 Thyristor-switched capacitor (TSC)

2.32.6.5 Mechanically switched capacitor

Capacitor is switched by circuit breaker. It aims at compensating steady-state reactive power. It is switched only a few times a day. It is very useful controller. However, with the advancement of power electronic switching devices, its use is gradually reducing.

2.32.6.6 Thyristor-controlled reactor/thyristor-switched capacitor

In TCR/TSC, both TCR and TSC are used. Such unit offers optimum solution in reactive power (VAR) compensation. Advantages of TCR and TSC are present along with wide range of zone of operation. During large disturbances or fault, performance of such unit is good.

Main features of TCR/TSC

1. Wide range of operating zone.
2. Control is continuous.
3. Transients occurred are very small and practically negligible.
4. Harmonic can easily be eliminated by the unit itself using it as filter or by adding some filter.
5. Losses are small.
6. Flexibility and redundancy are very high.

Control system of SVC

Control system of SVC unit aims to control the voltage needed at the point of power system where it is connected, and hence, it keeps system voltage close to the desired value. Continuously system voltage is measured and compared with set value (reference voltage). If any discrepancy between these two values is found, controller generates signal to change its voltage until equilibrium is achieved.

Controller's signal controls the firing of thyristor of TCR, TSC units, etc. Control units also supervise currents and voltages of different branches. In some cases, it also provides some sort of protection.

Thyristor firing and cooling system

Thyristors are fired electrically. Energy is normally taken from snubber circuit. Firing signals are sometimes transmitted through optical line. Thyristors are placed among heat sink. Heat sinks are kept in touch with cooling medium. Water, air, glycol, etc. are used as cooling medium.

Advantages of SVC in power transmission

Main advantages of using SVC in power transmission are as follows:

1. SVC is capable of stabilizing voltage of weak system.
2. Use of SVC in power transmission reduces transmission loss.
3. Use SVC in power transmission increases power transmission capacity. Thus, it decreases need of new lines with the increase of demand to some extent.
4. Use of SVC in power transmission increases the transient stability limit of the overall system.
5. SVC helps in damping small disturbances.
6. SVC helps in reducing power oscillation.
7. Better voltage control and stability are obtained with the application of SVC in power transmission.

Advantages of using SVC in power distribution

Main advantages of using SVC in power distribution are as follows:

1. Application of SVC in power distribution helps in stabilizing receiving-end voltage of LTL.
2. It improves stabilized voltage profile and thus increases capacity of power utility of the distribution system.
3. It reduces net reactive power compensation.
4. It reduces net loss in distribution.
5. It improves tariff.
6. It reduces voltage stress on machinery and better use of equipment is achieved.
7. It reduces voltage fluctuation.
8. It reduces flickering.
9. It reduces harmonics and harmonics distortion.

2.32.6.7 Static synchronous compensator

STATCOM acts as a source of reactive power to a system. Previously it was known as static condenser (STATCON). It consists of power source converter connected with capacitor. The converter is either voltage source converter (VSC) or current source converter (CSC). Based on the type of converter, STATCOM is classified into two categories: VSC-based STATCOM and CSC-based STATCOM. The synchronized voltage output is stepped up and fed to the system. Note that, it does not consist any active source. A VSC STATCOM is shown in Figure 2.56.

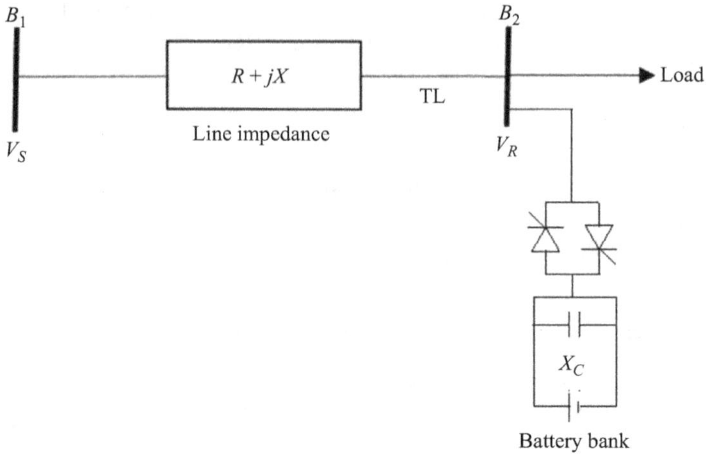

Figure 2.56 VSC-based STATCOM

2.32.6.8 Static synchronous generator

Static synchronous generator (SSG) also acts as a source of reactive power to the system. Unlike synchronous generator, SSG has no rotating part. It is almost same as STATCOM device; here capacitor unit is connected with DC source. As DC source, in some application, battery is used, and in some other applications, rectified DC is used. The synchronized voltage output is stepped up and fed to the system as shown in Figure 2.57.

2.32.7 Series–series controller

Here many series controllers are connected in series with transmission line in between two buses.

For single transmission line, two series controllers can be connected, and those controllers may be of same type and may not be of same type. Also for multiple number of transmission lines connected in parallel, same or different type of series controllers can be connected in series with the corresponding transmission line. Like series controller, series–series controllers also offer variable series impedance. Thus, line voltage drop accordingly be varied. When load varies, current changes and drop varies, which can be changed as desired for better flexibility. Depending upon requirement, series controller changes voltage drop and may deliver reactive power to the system or absorb reactive power from the system. When voltage is in quadrature with phase of current, only reactive power is involved in operation in terms of delivering or absorption of reactive power. When voltage is not in quadrature with phase of current, only reactive power is involved in operation in terms of delivering or absorption of reactive power along with some real or active power. Series–series controller is costlier than series controller.

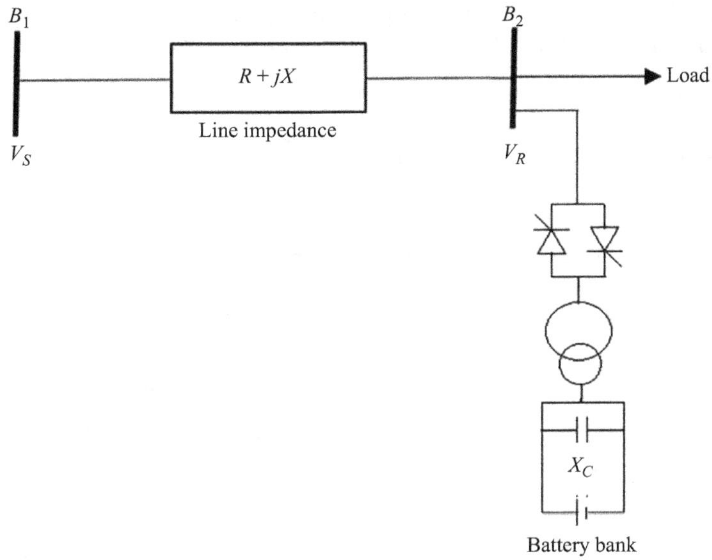

Figure 2.57 Static synchronous generator (SSG)

Main features of series–series controller

Main features of series–series controller are as follows:

1. It offers variable series impedance.
2. It offers variable series voltage drop.
3. It delivers reactive power to the system or absorbs reactive power from the system depending upon the requirement.

Demerits of series–series controller

When a bus is connected to more than two numbers of buses, for each line, separate series controller is needed; common series controller cannot be used.

2.32.8 Series–shunt controller

Here series controller is connected in series with transmission line in between two buses. Shunt controller is connected to a bus in parallel with transmission line. Sometimes series unit and shunt unit operate independently; sometimes their operations are interrelated. Thus by series-shunt controller, the advantages of using both series and shunt controller can be obtained.

Main series–shunt controllers used as FACTS device is Unified power flow controller (UPFC).

Unified power flow controller

Unified power flow controller (UPFC) consists of one shunt controller unit and one series controller unit. In some scheme, these two controllers operate independently, and in some applications, their operations are correlated. In most of the applications, STATCOM is used as shunt controller and SSSC is used as series controller.

2.33 Static phase shifter

Static phase shifter (SPS) is used to control the phase angle of voltages. Phase angle controller is placed in series with the transmission line. SPS is mainly one type of voltage controller which is capable of changing the phase angle of output voltage. It received sending-end voltage and considers angle of sending-end voltage as reference. It then introduces angle and injects voltage in the transmission line. It controls the phase of injected voltage and hence performs the phase shifting of receiving-end voltage.

2.34 FACTS and solar–wind hybrid grid

Modern day's grid systems are interconnected with various types of sources. Sources include conventional source like wind power plant. For such a grid system, issues like voltage fluctuation, deviation of frequency and power quality deterioration are very sensitive issues, and good care must be given over those issues. Application of FACTS devices like SVC, STATCOM and series capacitive reactance in grid improves stability and increases power transmission capacity over the existing line. With adding any new lines, nonconventional power like wind power, solar power can be added to the system. 'Dynamic energy storage' device is one of the new members in FACTS. Introduction of this device in the grid system adds some extra features to the grid like backup for renewable generation, better frequency regulation and renewable energy storage which can be utilized during high demand period. Solar photovoltaic resources contribute unidirectional power, whereas wind plant contribute with wide frequency variation. Therefore, proper converter and controller requires for connection with conventional grid system. In addition, special attention is needed for reactive power compensation for stable operation.

Special need of better reactive power regulation in a wind power plant

Wind power plant generates electric power from wind using induction generation or asynchronous generator. In normal practice, control of reactive power for asynchronous generators is improved by using doubly fed rotor induction generator (DFIG). During steady-state condition, its performance is good. However, during transient condition, it becomes very difficult to keep the operation within stable zone and to protect the system against overload. In fact, during transient condition, DFIG behaves like common induction generator. This demands dynamic control of reactive power at the point of common coupling (PCC).

Secondly, off-shore wind generation involves a lot of sea cable network which demands for better reactive power control for the system.

Other than connection, modern FACTS devices can be classified into three following groups:

1. Series devices
2. Shunt devices
3. Dynamic storage devices

In grid system, SVC is replaced by SVC with VSC or STATCOM and is connected at PCC as shown next.

Voltage-sourced converter
VSC is a widely used device in reactive power control system. It generates three-phase alternating voltage of controllable magnitude, phase and frequency. Its operation follows four quadrant space in *P*–*Q* plane. Here, active power (*P*) is used along the real axis and reactive power (*Q*) is used along the imaginary axis. In first quadrant mode of operation, VSC acts as reactive power generator as well as inverter. In second quadrant mode of operation, VSC acts as rectifier and can supply reactive power. In third quadrant mode of operation, application of VSC is rare. In fourth quadrant mode of operation, it absorbs both active and reactive power.

Advantages

- SVC with VSC replaces passive reactive components by electronics converter and reduces area requirement.
- It reduces work and commissioning cost and time as it is a factory-assembled unit.
- Natural relocate ability.
- Compact design.
- Low harmonic interaction with the grid.

STATCOM with SVC
The STATCOM is one of the key FACTS devices. Based on a voltage sources converter, the STATCOM regulates system voltage by absorbing or generating reactive power. Contrary to a thyristor-based SVC, STATCOM output current (inductive or capacitive) can be controlled independently.

2.35 Line capability

Capacity of transmission line refers to the maximum amount of power that can be transmitted through transmission line considering all loss components and maintaining electro-mechanical-thermal stability of the power system network. It is proportional to square of the line voltage and inversely proportional to line impedance.

2.36 Summary

In this chapter, fundamentals of power system relevant to the overhead lines have been learnt. Line parameters, resistance, inductance and capacitance are distributed in nature. They are passive, linear and bilateral; therefore, transmission and distribution lines have been modelled by two-port network modelling method followed by analysis. Different sequence impedances have been presented. Voltage regulation has been explained. Reactive power and voltage control techniques have been presented.

Further reading

[1] W. D. Stevenson Jr., Elements of Power System Analysis (4th Edition). McGraw-Hill, London; 2014.

[2] J. Arrillaga and B. Smith, AC-DC Power System Analysis. IET, Stevenage, United Kingdom; 1998, doi:10.1049/PBPO027E.

[3] J. M. Adams, Electrical Safety: A Guide to Causes and Prevention of Hazards. IET, Stevenage, United Kingdom; 1994, doi:10.1049/PBPO019E.

[4] E. Lakervi and E. J. Holmes, Electricity Distribution Network Design (2nd Edition). IET, Stevenage, United Kingdom; 2003, doi:10.1049/PBPO021E.

[5] S. Chattopadhyay, M. Mitra, and S. Sengupta, Electric Power Quality. Springer, New York; 2011.

[6] Y. H. Song and A. T. Johns, Flexible AC Transmission Systems (FACTS). IET, Stevenage, United Kingdom; 1999, doi:10.1049/PBPO030E.

[7] J. Arrillaga, High Voltage Direct Current Transmission (2nd Edition). IET, Stevenage, United Kingdom; 1998, doi:10.1049/PBPO029E.

[8] K. R. Padiyar and A. M. Kulkarni, Modeling and Analysis of FACTS and HVDC Controllers in Dynamics and Control of Electric Transmission and Microgrids. IEEE, Chichester, West Sussex, UK; 2019, pp. 145–194, https://doi.org/10.1002/9781119173410.ch5.

[9] G. G. Karady and K. E. Holbert, Electric Power Systems in Electrical Energy Conversion and Transport: An Interactive Computer-Based Approach. IEEE, Chichester, West Sussex, UK; 2013, pp. 1–29, doi:10.1002/9781118498057.ch1.

[10] H. M. Ryan, High Voltage Engineering Testing (3rd Edition). IET, Stevenage, United Kingdom; 2013, doi:10.1049/PBPO066E.

Chapter 3
Line support, foundation and mechanical sag

This chapter presents different aspects of line support, foundation and mechanical sag of transmission line. It starts with introduction and then describes the design aspects with respect to distribution system and then with respect to transmission system. Components of overhead lines have been described. Different types of line supports have been presented. Different feature of wooden pole, steel pole, concrete pole and lattice tower have been presented. Foundation methods have been presented for different line support. Classification of foundation has been made. Different types of monoblock and compact foundations have been presented. Tests for supports have been highlighted. Mechanical sag with tension for overhead lines has been presented. Effects of ice and wind on sag have been described.

3.1 Introduction

Overhead power transmission lines are made of line support, conductor and line insulator. Support is a mechanical structure used to hold the conductors with certain clearance from the ground. Conductor is the main part of overhead line used to carry electric power. Line insulator is the remaining important part of overhead transmission line located in between line support and conductor. Ground clearance depends on the operating voltage of the transmission system whereas the dimension of the line conductor depends on the transmission capacity of active power through the conductors. With the increase of voltage level, more ground clearance is required, and with the increase of load or active power transmission capacity, number, weight and dimension of conductor increase. These two aspects require bigger and higher line support to provide more mechanical strength and ground clearance. On the other hand, due to the weight of the conductors, lowering of conductor occurs along the line, which is commonly known as sag. Different types of line support and mechanical sag have been discussed in the following sections.

3.2 Components of overhead lines

Overhead lines are installed for the purpose of transmitting electric power from one location to another through conductors. Line conductors are positioned above ground levels. Overhead lines are exposed to different types of electrical and mechanical failure which needs protection and safety measure. Therefore, major components of transmission line may be listed as follows:

1. Line support
2. Conductor
3. Insulator
4. Protective devices

3.3 Design aspects of distribution system

3.3.1 Basic consideration

Line support design for distribution system takes care of the following aspects: voltage level of distribution, number of conductors and number of phases (single or three). Based on electrical parameters, type of pole to be used in power distribution is chosen. In addition, location is also important with respect to cost estimation as well as minimum ground clearance requirement. Voltage level of distribution system is kept low in urban areas. Based on voltage levels, height and span of the poles are decided.

3.3.2 Classification of line support used in distribution system

Line supports used in distribution system are divided into two main categories:

1. Pole
2. Steel tower

Poles are divided into following three main categories:

1. Wooden pole
2. Concrete pole
3. Steel pole

Steel towers are divided into the following categories:

1. I-shaped – again is of two types, viz. narrow based and broad based (common)
2. H-shaped

3.3.2.1 Wooden pole

Woods are available in plenty in nature. Long dry-wooden columns are used as poles. Wooden poles are normally used where they are available. If the transportation of wooden poles is needed, in most of the cases, it is done by river, wherever available.

Application

Wooden poles are used in low-voltage (LV) or low-tension (LT) lines. Wide applications of wooden poles are observed up to 11 kV. However, they find application up to 40 kV. In rare cases, double-pole wooden structures are used up to 220 kV.

Advantages and disadvantages

Advantages

Main advantages of wooden poles are:

1. Wood is naturally available
2. Cost of wooden pole is low
3. Weight is low
4. Dry wood is a good insulator
5. Manufacturing cost is low
6. Cost of transportation is low

Disadvantages

Main disadvantages of wooden poles are:

1. Height is limited, which cannot be increased.
2. Long wooden structures are not available everywhere.
3. As height is limited, for high-voltage AC (HVAC) or extra-HVAC (EHVAC) transmission, wooden poles cannot be used.

Selection features

The following features are considered to use wood as a pole:

- Height: Length of the wood should be moderately high.
- Straightness: Wood to be used as a pole should be straight.
- Fungus free: Wood should be fungus free. It ensures long durability of the pole.
- Seasoned: Before being used as a pole, wood must be well seasoned for at least 6–12-months durations.

In different parts of the world, different types of tree are found available that are suitable for processing and manufacturing of poles. Some of them are given in the following:

- America: pine, oak, chestnut, cedar, cypress, juniper, tamarack, fir, redwood.
- Europe: Baltic pine, Norway pine, Douglas fir, cypress, yellow pine, oak, chestnut.
- Asia Pacific: redwood, pine, etc.

Classification of wooden pole

Based on the shape of a pole, wooden poles may broadly be classified as follows:

1. I-shaped
2. A-shaped
3. H-shaped

The following types of conductor positions are observed in wooden poles:

1. Vertical I for single phase.
2. Vertical I-shape for three-phase four-wire system.
3. Symmetrical triangular shape for three-phase three-wire system.
4. Unsymmetrical triangular for three-phase three-wire system.
5. Horizontal for three-phase three-wire system.

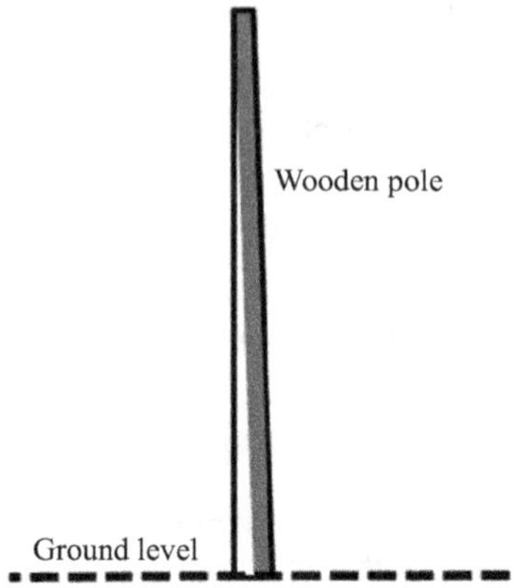

Figure 3.1 'I'-shaped wooden pole

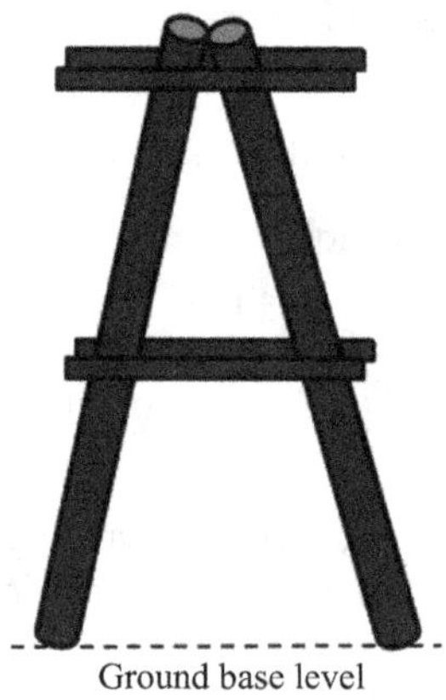

Figure 3.2 'A'-shaped wooden pole

I-shaped wooden pole: I-shaped wooden pole has been shown in Figure 3.1.

A-shaped wooden pole: A-shaped wooden pole has been shown in Figure 3.2.

H-shaped wooden pole: H-shaped wooden pole has been shown in Figure 3.3.

Wooden pole carrying transformer: Sometimes wooden poles are used to carry small distribution transformers. For this purpose, double-pole H-structure is more suitable. A sample transformer mounted on a wooden pole has been shown in Figure 3.4. The transformer is mounted at cross arm near middle of the height of the pole.

Height and diameter

Height of a wooden pole cannot be increased beyond availability in the type of wood. The diameter of the wooden pole is approximately uniform. Diameter at the top is slightly less than diameter at the ground level as shown in Figure 3.5. With the increase of height, diameter at the ground level increases.

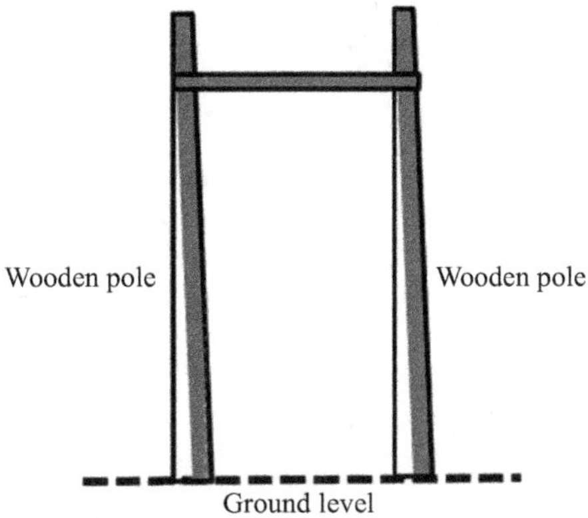

Figure 3.3 'H'-shaped wooden pole

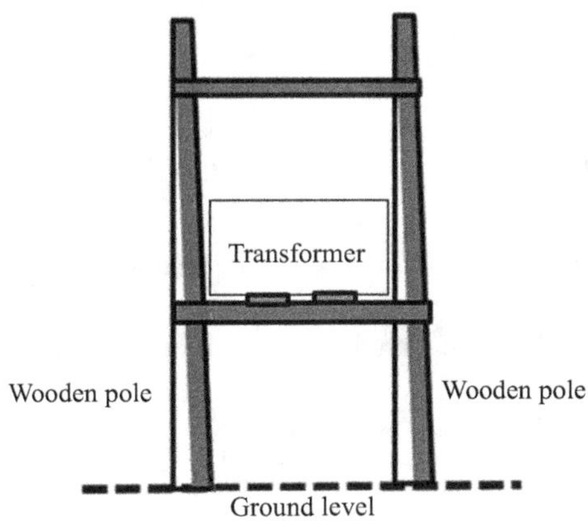

Figure 3.4 Position of distribution transformer carried by double wooden poles

Let,

d = Diameter at the top
D = Diameter at the ground level; $D > d$
H = Height of the top from the ground level

$$D = d + tH \tag{3.1}$$

t is known as 'taper'.

$$tH = D - d \tag{3.2}$$

$$\text{or,}\ t = \frac{D - d}{H} \tag{3.3}$$

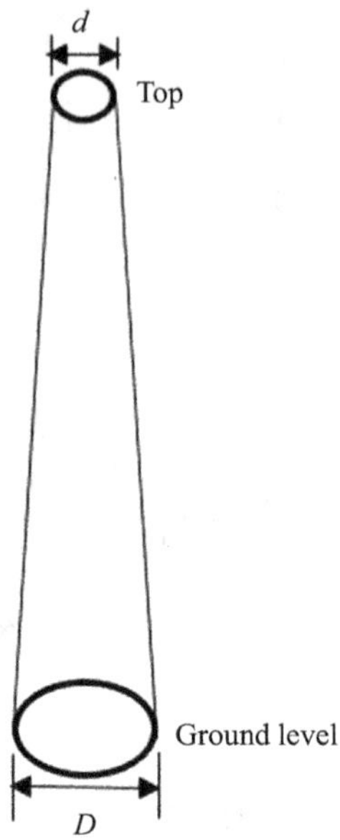

Figure 3.5 Single wooden pole structure

Thus, tapper is the ratio of difference in diameter of top from bottom and height of the pole.

Volume of wood of a pole may be expressed as follows:

$$\text{Volume} = \frac{1}{3}H \times \frac{\pi}{4}\left(D^2 + Dd + d^2\right) = \frac{\pi H}{12}\left(3d^2 + 3dtH + t^2H^2\right)$$

Wood strength

Strength of the wooden pole depends on the following factors:

- *Pull by the conductor:* Pulling force by the conductors plays important role in the calculation of stress and strength of a wooden pole. If poles are located at equal span in linear way, pulls by conductor from both sides of it balance each other and do not have effect on stress calculation. But for unequal spans, resultant pull has effect on stress or strength calculation.

  If the pole is located at a bending path, pull is adjusted by guy wire. Net pull by conductor and guy wire is to be considered.
- *Ice effect:* Ice on the conductor may change the pull force.
- *Wind effect:* Wind flow changes pull force. Sometimes the conductors get stuck together due to heavy wind pressure that may change significant pull force.
- *Dead weight of the pole and conductor:* Dead weight of the pole and conductor acts downwards, and if neglected in strength, calculation of poles is located in short and medium span.

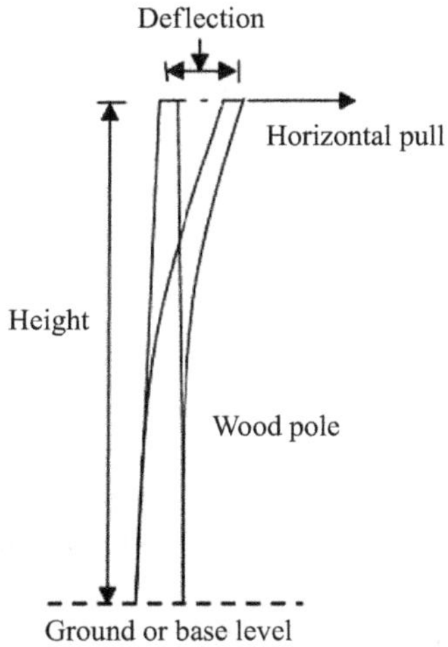

Figure 3.6 Single wooden pole: pull direction

Let, for the wooden pole shown in Figure 3.6,

d = Diameter at the top
D = Diameter at the ground level
H = Height of the top from the ground level
t = Taper
x = Depth from the top where stress will be calculated
D_x = Diameter at depth x

$$\therefore D_x = d + tx \tag{3.4}$$

P = Net pull acting horizontally
Moment of inertia at remote end fibre at depth $x = Px$

Also,

T_x = Stress at moment of inertia at remote end fibre at depth x
Z = Section factor

Moment of inertia at depth x can be written as follows:

Stress at moment of inertia at remote end fibre at depth x

$\times$ section factor

$= T_x \times Z$

$$Z = \frac{\pi D_x{}^3}{32} \tag{3.5}$$

$$\therefore\ Px = T_x \frac{\pi D_x{}^3}{32} \tag{3.6}$$

$$\therefore\ \text{Stress at depth } x = T_x = \frac{32Px}{\pi D_x{}^3} = \frac{32Px}{\pi(d + tx)^3} \tag{3.7}$$

$$\text{Stress at depth } H, \text{i.e. at bottom} = T_H = \frac{32PH}{\pi D_x{}^3} = \frac{32PH}{\pi(d + tH)^3} \tag{3.8}$$

Maximum breaking possibility

Diameter of the wooden pole increases as depth from the top increases. Stress at depth x may be calculated from (3.7).

For maximum stress, differential of stress will be zero. Thus,

$$\frac{dT_x}{dx} = 0 \tag{3.9}$$

This results as follows:

$$x = \frac{d}{2t} \tag{3.10}$$

$$\therefore\ D_x = d + t \times \frac{d}{2t} = 1.5d \tag{3.11}$$

Breaking possibility reaches maximum at a depth where diameter becomes equal or more than 1.5 times of the diameter found at the top (i.e. $D \geq 1.5d$).

$$Px = T_x \frac{\pi D_x{}^3}{32} \tag{3.12}$$

$$P = T_x \frac{\pi D_x{}^3}{32x} = T_x \frac{\pi(1.5d)^3}{32 \times \frac{d}{2t}} = 0.662 \times d^2 t T_x \tag{3.13}$$

$$d = 1.229 \times \sqrt{\frac{P}{tT_x}} \tag{3.14}$$

Deflection

Deflection of a wooden pole is calculated by pull, moment of inertia and Young's modulus of the wood.

Let,

P = Net pull acting horizontally
Y = Young's modulus
I = Moment of inertia

The wooden pole is fixed at the bottom and facing horizontal pull at top. This can be compared with a beam fixed at one end and facing force at another end. In such cases, deflection may be written as

$$x_{defl} = \frac{1}{3}\frac{PH^3}{YI} \tag{3.15}$$

If including wind effect, pull is uniformly distributed, deflection will be

$$x_{defl} = \frac{1}{8}\frac{PH^3}{YI} \tag{3.16}$$

If the diameter of the wooden pole is uniform, the moment of inertia may be written as

$$I = \frac{\pi d^4}{64} \tag{3.17}$$

If diameter is not uniform, i.e. if diameter is different at top (d) and bottom (D_H), moment of inertia will be

$$I = \frac{\pi d {D_H}^3}{64} \tag{3.18}$$

From the previous equation, deflection (x_{defl}) of the top is seen as

1. proportional to horizontal pull (shown in Figure 3.7).
2. proportional to the cube of the height (shown in Figure 3.8).

Maintenance and life time

The average expected lifespan is 30 years. In reality, lifespan varies from 20 to 35 years. Main cause of deterioration of a wooden pole is fungus. Fungus comes into action and survives because of soil nature, wood cell and climatic condition. Care is taken for installation, preservation and maintenance. Time of cutting the timber is very important. Absorption of preservative oils depends on the time of cutting. Autumn-cut pole is easier to treat with preservatives than summer-cut poles. It may

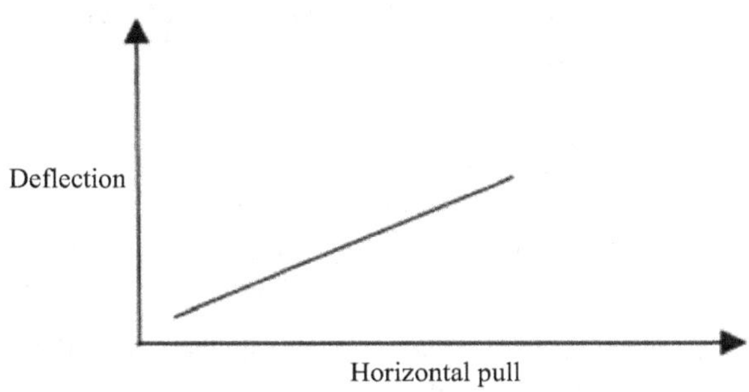

Figure 3.7 Deflection of single wooden pole top versus horizontal pull

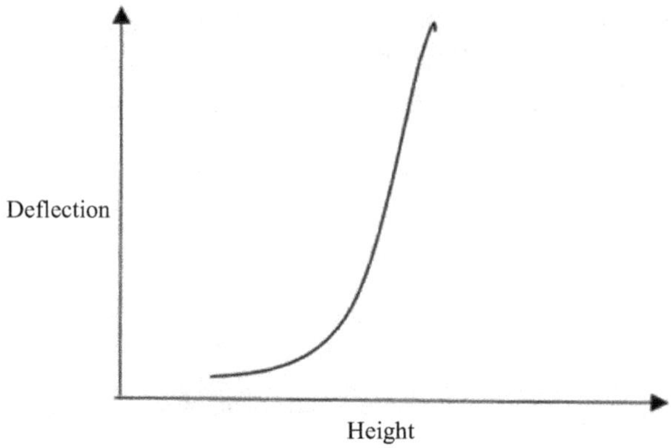

Figure 3.8 Deflection of single wooden pole top versus height

be noted that only living object like fungus can destroy wood in the presence of sufficient light, air and moisture; on the other hand, full dry condition or water immersed condition shall not destroy wood significantly. Various chemicals and oils are used as preservative. Coal tar creosote oil is widely used for this purpose. The oil may be applied in different ways. Most popular three methods are high-pressure treatment using Bethel system, open-tank treatment and brush treatment.

3.3.2.2 Concrete pole

Concrete poles are manufactured by concrete materials (Figure 3.9). Steel rods are reinforced in the vertical direction, and steel rings are reinforced in the horizontal direction.

Cross section

Cross-sectional area at ground level is slightly greater than the cross-sectional area at top depending upon the tapper. The following types of cross sections are found for a concrete pole:

Solid square

Solid square shape of concrete pole is very common. Their mechanical strength becomes very high; weight of the pole also increases. Pole with solid square cross section has been shown in Figure 3.10(a).

Solid rectangular

Solid rectangular shape of a concrete pole is very common. Its mechanical strength becomes very high; weight of the pole also increases. Pole with solid rectangular cross section has been shown in Figure 3.10(b).

Hollow circular

In this type of pole, cross section is made circular in shape and the inner part of the cross section is made hollow to reduce mechanical weight and also cost of concrete material. Hollow circular cross section has been shown in Figure 3.10(c).

Figure 3.9 Concrete pole

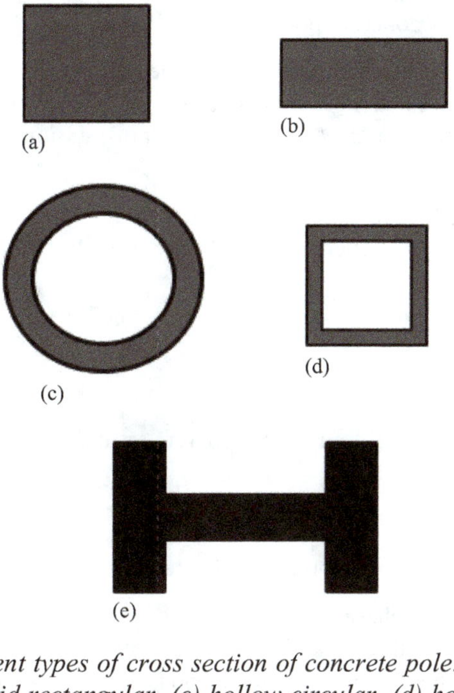

Figure 3.10 Different types of cross section of concrete pole: (a) solid square, (b) solid rectangular, (c) hollow circular, (d) hollow square and (e) H or two T

Hollow square

In this type of pole, cross section is made square in shape and the inner part of the cross section is made hollow of square shape to reduce mechanical weight and also cost of concrete materials. Hollow circular cross section has been shown in Figure 3.10(d).

H or two (or, double) T shape

In this type of pole, the cross section looks like H or two T connected in the reversed direction. This reduces mechanical weight and also the cost of concrete materials. Comparatively this type of pole is easier to manufacture than hollow circular or square cross section. H or two T cross section has been shown in Figure 3.10(e).

Application

Concrete poles are used in LV or LT lines.

Advantages and disadvantages

Advantages

Advantages of concrete poles are as follows:

1. Mechanical strength is high.
2. Concrete poles can be manufactured easily.
3. Manufacturing cost of concrete pole is lower than steel pole and steel tower.

Disadvantages

Main disadvantages of concrete poles are as follows:

1. Weight is high and transportation is not cheap.
2. Single concrete pole cannot be made too high.
3. As height is limited, for HVAC or EHVAC transmission, concrete poles cannot be used.

Classification of concrete pole based on shape

Based on the shape of the concrete pole, it may broadly be classified as follows:

1. I-shaped
2. H-shaped
3. A-shaped
4. Four-pole structure

I-, A-, H-shaped and four-pole concrete structures have been shown in Figure 3.11.

Conductor position

The following types of conductor positions are observed in a concrete pole:

1. Vertical I for single phase
2. Vertical I-shape for three-phase four-wire system
3. Symmetrical triangular shape for three-phase three-wire system
4. Unsymmetrical triangular for three-phase three-wire system
5. Horizontal for three-phase three-wire system

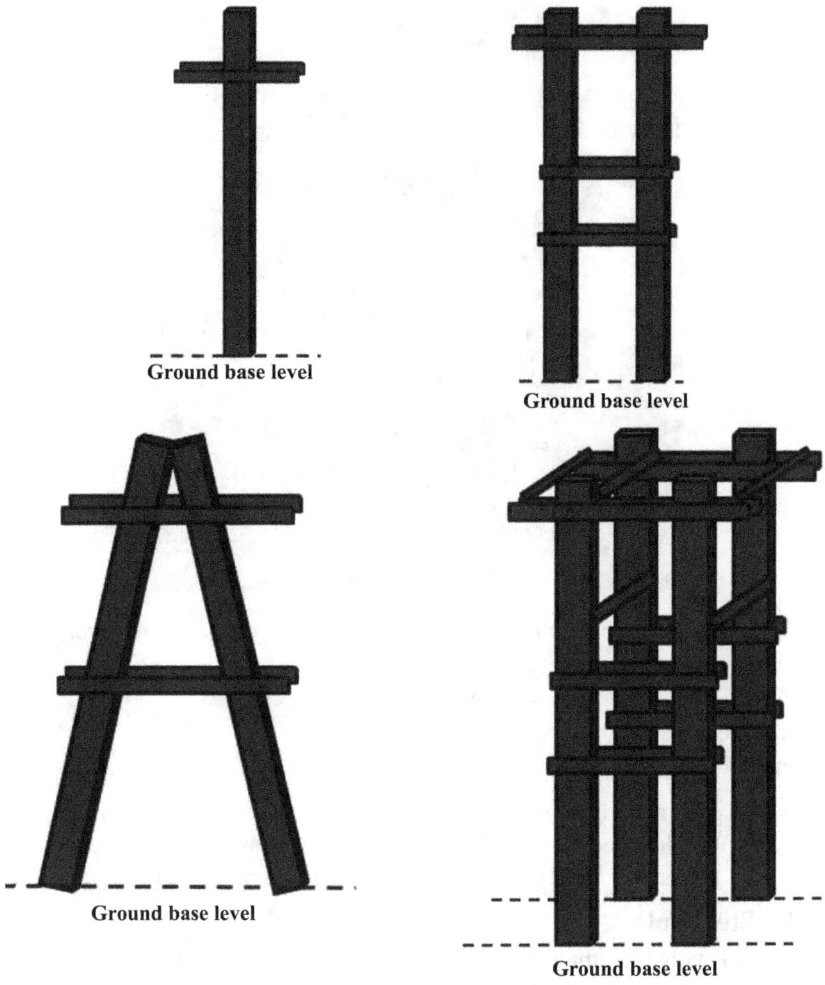

Figure 3.11 Different shapes of structure with a concrete pole

Concrete pole carrying a transformer

Sometimes concrete poles are used to carry small distribution transformers. For this purpose, double-pole H-structure is more suitable. A sample transformer mounted on a concrete pole has been shown in Figure 3.12. The transformer is mounted over the cross arms near the middle of the height of the pole.

Tilt

Due to wind pressure and unequal pull by the conductors, the concrete pole has tendency to tilt. It depends on

Figure 3.12 Double concrete poles carrying a distribution transformer

1. Soil strength
2. Wind pressure
3. Conductor pull

A tilted pole and tilted pin type insulator due to a 130 km/h wind speed of cyclone Amphan has been shown in Figure 3.13 and Figure 3.14 respectively. To prevent total fall, a guy wire is connected to avoid complete falling of the pole.

3.3.2.3 Steel pole

This type of poles is manufactured by steel (Figure 3.15). It provides high mechanical stress. Steel pole offers a greater length of pole than a concrete pole.

Cross section

Cross-sectional area at ground level is slightly greater than the cross-sectional area at top depending upon the tapper. The following types of cross sections are found for a concrete pole:

Hollow circular

In this type of pole, the cross section is made circular in shape and the inner part of the cross section is made hollow to reduce mechanical weight and also cost of concrete material. Hollow circular cross section has been shown in Figure 3.16(a).

Hollow square (very rare)

In this type of poles, cross section is made square and the inner part of the cross section is made hollow of square shape to reduce mechanical weight and also cost of concrete material. Hollow circular cross section has been shown in Figure 3.16(b).

Figure 3.13 *Concrete pole tilted due to wind pressure (Amphan, West Bengal, India, 2020)*

Figure 3.14 *Pin insulator axis tilted due to wind pressure (Amphan, West Bengal, India, 2020)*

Figure 3.15 *Steel poles*

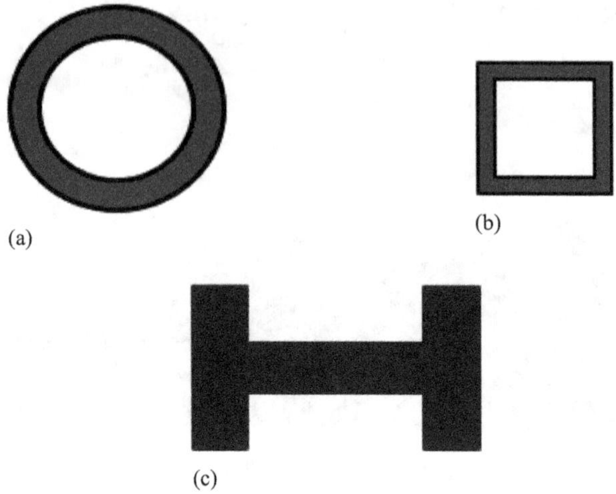

Figure 3.16 Different types of cross section of a concrete pole: (a) hollow circular, (b) hollow square and (c) H or two T

H or two (or, double) T-shape

In this type of poles, cross section looks like H or two T connected in the reversed direction. This reduces mechanical weight and also cost of concrete materials. Comparatively this type of pole is easier to manufacture than hollow circular or square cross section. H or two T cross section has been shown in Figure 3.16(c).

Application

As steel poles can be made of higher length, they are used in LV or LT lines as well as low-medium voltage lines also.

Advantages and disadvantages

Advantages

Advantages of steel poles are as follows:

1. Mechanical strength is high.
2. Steel poles can be manufactured easily.
3. Manufacturing cost of steel pole is lower than steel tower.

Disadvantages

Main disadvantages of steel poles are as follows:

1. Weight is high and transportation is not cheap.
2. Single steel pole cannot be made too high.
3. As height is limited, for HVAC or EHVAC transmission, steel structure is not suitable and preferable.

Classification of a steel pole

Structural shapes of steel poles are the same as a concrete pole. However, greater height can be obtained by steel pole structures. Based on the shape of the steel, a pole may broadly be classified as follows:

1. I-shaped
2. H-shaped
3. Four-pole structure

I-shaped steel pole
I-shaped steel pole has been shown in Figure 3.17.

H-shaped steel pole
H-shaped steel pole has been shown in Figure 3.18.

Four-pole structure
Four-pole structure has been shown in Figure 3.19.

Steel pole with multiple circuits
Sometimes steel poles are seen to carry multiple circuits; a sample has been shown in Figure 3.20.

Steel pole carrying a transformer
Sometimes steel poles are used to carry small distribution transformers. For this purpose, double-pole H-structure is more suitable. The transformer is mounted at cross arm near middle of the height of the pole. It is placed with safety measures, protective units and bus systems.

3.3.3 Conductor positions for pole support

The following types of conductor positions are observed in pole supports (wood, concrete and steel types) (Figures 3.21–3.30):

1. Vertical I for single phase
2. Vertical I-shape for three-phase four-wire system
3. Symmetrical triangular shape for three-phase three-wire system
4. Unsymmetrical triangular for three-phase three-wire system
5. Horizontal for three-phase three-wire system

Figure 3.17 'I'-shaped steel pole under construction

Figure 3.18 'H'-shaped steel pole

Figure 3.19 Four-pole steel structure at bend path of 33 kV distribution line

3.3.4 Guard wire in distribution system

Guard wires are placed below the phase conductors. Normally, guard wires are provided with the towers while transmission line crosses roads. Guard wires are directly connected with the tower, and they are grounded through the tower body (Figure 3.31).

Figure 3.20 Steel pole carrying multiple circuits

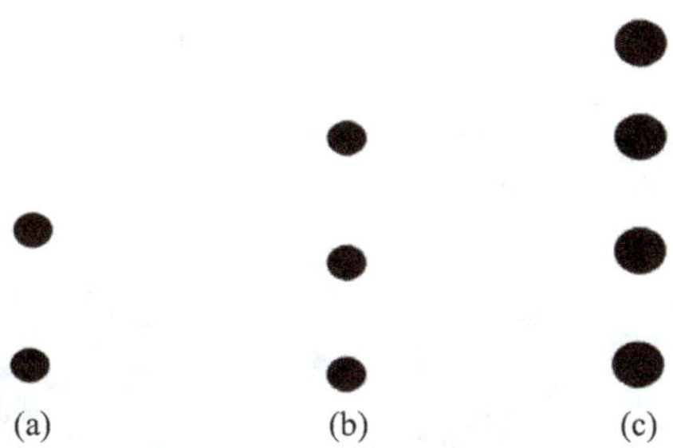

Figure 3.21 Conductor position for distribution system with vertical 'I'-shape: (a) two wires, (b) three wires, (c) four wires

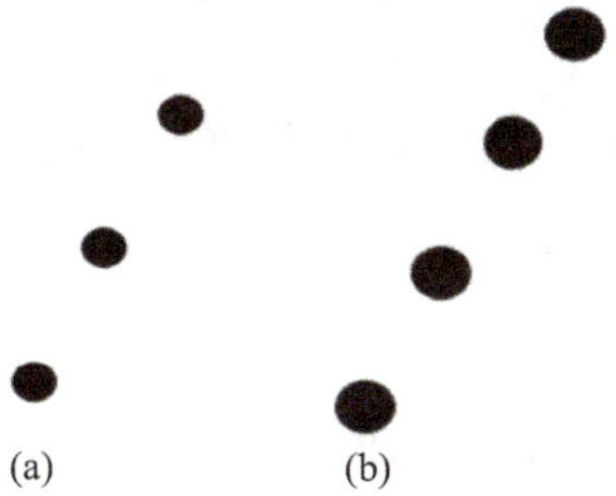

Figure 3.22 Conductor position with tilted 'I'-shape: (a) three wires, (b) four wires

Figure 3.23 Horizontal linear conductor position with three-wire system

Figure 3.24 Triangular conductor position for three-wire distribution system: (a) symmetrical, (b) unsymmetrical

Figure 3.25 'I'-shaped conductor position for three-wire double-circuit distribution system

Figure 3.26 Horizontal linear conductor position with three-wire 11 kV distribution system

3.3.5 Guy in distribution system

Guys are additional steel wires provided to give additional mechanical support to the pole or tower. Based on the type of tower and soil strength, one or multiple number of guy wires are provided (Figures 3.32 and 3.33).

Figure 3.27 'I'-shaped conductor position for four wires 440 V distribution system

Figure 3.28 'I'-shaped conductor position for three-wire 11 kV distribution system

3.3.6 Tower

Towers are used from medium to extra-high voltages. They are made of steels, and height can be made as required. Additional advantage of tower is that they are made by assembling small steel rod/structures at the spot/site; this reduces transport cost and time to a great extent. For example, only 3–4 days are required to make the structure of a tower from the ground level for 400 kV lines.

Figure 3.29 *'I'-shaped conductor position for three-wire 33 kV double-circuit distribution system*

Figure 3.30 *'Symmetrical triangular'-shaped conductor position for three-wire 33 kV distribution system*

Figure 3.31 *Connection of guard wire below the lines*

(a) (b)

Figure 3.32 (a) Four-guy double pole of 33 kV short transmission line and (b) jointing of guy with poles

Figure 3.33 Jointing of guy with a clamp of 11 kV steel poles

3.3.6.1 I-shaped tower

Depending upon the base area of the tower, I-shaped tower may again be divided into two categories: narrow based and broad based. Depending on the network connection, again it is divided into two categories: single circuit and double or multiple circuit. Normally, most of the single-circuit I-shaped towers are narrow-based towers. I-shaped towers are used in LV and medium voltage networks. Single-circuit and double-circuit I-shaped towers are shown in Figure 3.34.

Line conductors are connected through a string of overhead disc-type line insulators with the tower. The ground wire is placed at the top of the tower and is connected directly with the tower with any insulator.

3.3.7 Jumper in distribution system

Jumper is a conductor used to make connection to the transmission line wires connected to the insulator end (Figure 3.35). Normally, three categories of jumpers are seen:

1. Jumpers hanging below the insulator.
2. Jumpers passing by the side of the insulator and line wire. These types of jumpers are placed at the side of the main line either on separate insulator(s) or strained by or suspended below the string of insulators.

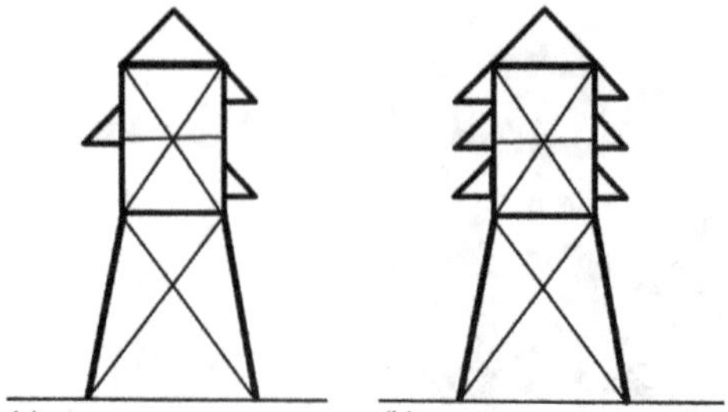

Figure 3.34 (a) Single and (b) double-circuit I-shaped tower

Figure 3.35 (a) Jumper passing above lines and (b) jumpers passing by side of the lines

3. Jumpers passing above the insulators. These types of jumpers are placed above the main line on separate insulator(s).

3.3.8 Foundation of distribution poles

Foundations of distribution poles are simple than lattice towers. Poles are buried after excavation. Around one-sixth of length of the pole is kept below ground level. It varies depending on the strength of soil and probable water percentage in the soil. Sometimes, concrete pad or slab is made at the bottom of foundation. After placing the pole in the foundation hole, vacant space is filled up by soil and stones; soil is pressed surrounding the poles areas. For wooden poles, protective measures are taken to eliminate or reduce the probability of natural damage of wood due to fungus in presence of mixture of air and water. Foundation area and soil pressing have been shown in Figures 3.36 and 3.37, respectively.

Figure 3.36 Foundation area for a concrete pole of 440 V distribution system

Figure 3.37 Soil pressing at foundation work of concrete pole for 440 V distribution system

3.4 Design aspects in transmission system

3.4.1 Line support used in transmission system

Towers are used as support in all types of transmission system like short, medium and long transmission lines. However, in short transmission line, poles are still used in many places. Poles have been discussed in detail in the previous section. Work of tower consists of foundation, erection and stringing. In this section, towers as support with foundation and stringing will be discussed in detail.

Towers are used for a wide range of voltages from medium to extra-high voltages. As discussed in towers for distribution system, they are made of steels and height can be made as required. They are installed by assembling small steel rod/ structures at the spot/site, reducing transport cost and time to a great extent. Few days (about 3–4 days) are required to make the structure of a tower from ground level for 400 kV lines.

3.4.2 Classification of tower

3.4.2.1 Classification based on acting force of conductor

Based on acting force of the conductor and connection of line wire with insulators, towers are divided into the following four categories:

1. Suspension tower
2. Angle tension tower
3. Angle tower
4. Dead-end tower

Suspension tower

In suspension towers, conductors are placed and suspended below the insulator strings. Suspension towers are used where transmission line goes straight in the linear direction, and there is no end or bend angle at the path of the transmission line.

Angle tension tower

In angle tension towers, conductors are strained with the insulator strings. Angle tension towers are used where there is bend angle at the path of the transmission line (Figure 3.38).

Angle tower

Angle tower is very rare type of angle tension towers. In angle towers, the conductors are strained with the insulator strings and are subjected to ice and wind pressure.

Dead-end tower

In dead-end towers, conductors are strained with the insulator strings. Angle tension towers are used where transmission line starts or ends near a substation.

(a)

(b)

Figure 3.38 (a) Angle tension lattice tower of a 765 kV transmission system (under construction); (b) angle tension lattice tower of a 132 kV transmission system

Table 3.1 Angles of deviation of towers of different classes

Class of tower	Angle of deviation (degree)
A	0–2
B	>2, <15
C	>15, <30
D	>30, <45

3.4.2.2 Classification based on the angle of deviation

Based on the angle of deviation, towers are divided into the following four categories:

1. Class A
2. Class B
3. Class C
4. Class D

The angles of deviation of above-mentioned classes have been given in Table 3.1.

3.4.3 Materials used in tower

Steel and aluminium are used in tower construction. In most of the cases, steel is used. Aluminium is used where light weight of the tower is desired. Aluminium towers are found in hilly regions. However, in most of the application, steel is used. It is galvanized before use.

3.4.4 Structure

Tube and lattice structures are used for support construction.

3.4.4.1 Tube structure

Tube structure is used in low and medium-voltage applications.

3.4.4.2 Lattice structure

Lattice structures are found in most of the tower construction. They are easy to assemble. Members of the tower are brought to the site where tower is made. They are then assembled during erection. The cost of lattice tower contributes 35–45% of the total cost of transmission line.

Few days (3–6 days) are required to build one tower. Tower height depends on the following major factors:

- Voltage level
- Permissible mechanical sag
- Required clearance of conductor from the ground level
- Force or pressure of conductor
- Wind and ice effect in the region

Height of the electrical lattice steel tower ranges from 10 m to around 640 m. Heights of some well-known towers are given in Table 3.2.

Table 3.2 Heights of some well-known lattice towers

Tower	Approximate height (m)
Petronius Compliant Tower, USA	640
Tokyo Skytree, Japan	634
Boldpate Compliant Tower, USA	581.5
Dragon Tower, China	336
St. Petersburg TV Tower, Russia	326
Eiffel Tower, France	324

Advantages of lattice tower

Lattice structures have the following advantages:

1. High mechanical strength to carry forces of conductor, wind, ice, etc.
2. Height can be made as high as required.
3. Members are brought to the site and then assembled. This reduces the cost of transportation as well as overall construction.
4. It can carry pull of bundled conductor easily.
5. It can withstand more tension or strain force of a conductor.
6. It can carry multiple circuits.
7. It can provide a multiple number of earth wires or sky wires as required for protection and safety purpose.
8. At the ground level, under-tower agriculture is possible.

3.4.5 Different parts of tower

Different parts of a lattice steel tower have been shown in Figure 3.39.

3.4.5.1 Ground level or base level

Ground level or base level refers to the level where foundation ends and tower height starts. Construction of tower legs starts from this level. Sample base level of 132 kV line has been shown in Figure 3.40.

3.4.5.2 Legs

Steel members starting from the base level are called main legs. Through four legs, the entire pressure is distributed and goes to ground. Legs are placed with some angle decided from the foundation. Angle and inter-distance of legs are decided by the height of the tower and pressure subjected to the tower. Legs of a tower of 765 kV transmission line have been shown in Figure 3.41.

3.4.5.3 Waist

End part of the main legs is known as waist. It is located in between cage and leg. Sometimes length is provided in waist.

3.4.5.4 Tower body

Tower body refers to the part of the tower above base level to the part below the cage (as shown in Figures 3.39 and 3.42).

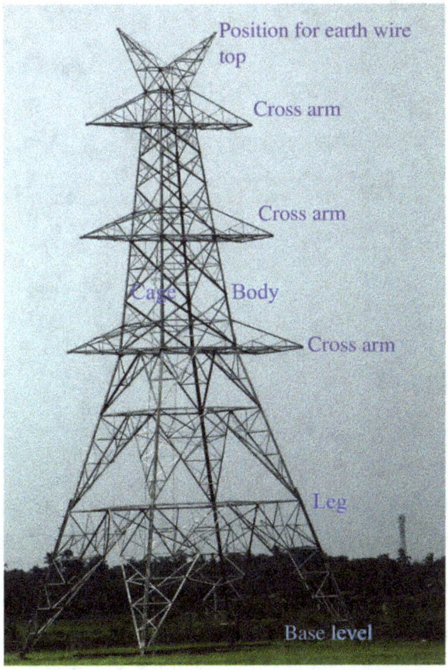

Figure 3.39 Lattice tower of a 765 kV transmission system

Figure 3.40 Ground level or base level of a tower of 765 kV transmission system

3.4.5.5 Cage

The structure placed above the waist and below the top is known as cage of the tower. It carries beam and cross arms (as shown in Figures 3.39 and 3.42).

Beam

Beams are the part of tower to hold the cross arms. Pressure of the conductor goes to tower through cross arm and beam. Normally, they are placed in parallel to the horizontal axis.

Cross arm

Cross arm holds the insulator or a string of insulators and conductors thereafter (as shown in Figures 3.39 and 3.42). Position, height and inter-distance of the conductors vary for different applications and conductor positions. Locations of end points of the cross arms may be as follows but not limited to

Figure 3.41 Legs of a tower of 765 kV transmission system

1. Linear horizontal
2. Linear vertical
3. Symmetrical triangular
4. Unsymmetrical triangular

The number of cross arms depends on the number of circuits carried by the tower.

In HVDC transmission, the number of cross arms depends on the nature of circuit whether monopolar or bipolar.

Top or peak

Top refers to the highest part of the tower located above cage structure (as shown in Figures 3.39 and 3.43). It is also known as the peak of the tower. At medium-voltage single-circuit lines, phase conductors are connected at top through a pin-type insulator. But at the top of the tower, an earth wire is connected directly with the tower.

Arms

Tower legs are supported and inter-connected by arms. There are two categories of arms:

Diagonal arm: Diagonally located with the vertical and horizontal axes of the tower.
Horizontal arm: Located in parallel with the horizontal axis.

Based on the presence of different types of arms, towers are categorized as follows:

1. All diagonal arms in one direction
2. Cross-linked diagonal arms
3. Diagonal arms coupled with horizontal arms

Figure 3.42 *Tower body, cage, cross arms and top for double-circuit transmission system*

Figure 3.43 *Tower top with provision for two earth wires or sky wires for double-circuit transmission system*

3.4.6 Guy

Guy is an additional steel wire provided to give additional mechanical support to the pole or tower. Based on the type of tower and soil strength, one or multiple number of guy wires are provided.

3.4.7 Earth wire or sky wire

Earth wires (EW) are connected directly to the tower without any insulator at top. As they are grounded through the tower, these wires are known as earth wire. Also as they are located at the top of all conductors below the sky, they are also commonly known as sky wire. The number of conductor depends on the voltage level of the line, the height of the tower and the number of circuits.

3.4.8 Jumper

Jumper is the conductor used to make connection to the transmission line wires connected to the insulator end. Normally, three categories of jumpers are seen:

1. Jumpers hanging below the insulator.
2. Jumpers passing by the side of the insulator and line wire. These types of jumpers are placed at the side of the main line either on separate insulator(s) or strained by or suspended below the string of insulators.
3. Jumpers passing above the insulators. These types of jumpers are placed above the main line on separate insulator(s).

Different types of jumpers will be discussed in Chapter 6.

3.4.9 Guard wire

Guard wires are placed below the phase conductors. Normally, guard wires are provided with the towers while transmission line crosses roads. Guard wires are directly connected with the tower, and they are grounded through the tower body.

Guard wire connection has been shown in Figure 3.44. Guard wire connected below 33 kV line has been shown in Figure 3.45.

3.4.10 Damper

Due to wind pressure, line conductors undergo continuous mechanical vibration. Dampers are used to reduce the vibration. Dampers are placed on the line conductors near the connection made with a string of insulators.

3.4.11 Clamp

Clamps are used to make strong mechanical holdings or connection in the following locations depending upon the requirement of the support:

- Between insulators and cross arms
- Between insulators
- Between insulators and conductors
- Between cross arm and support (for pole supports)

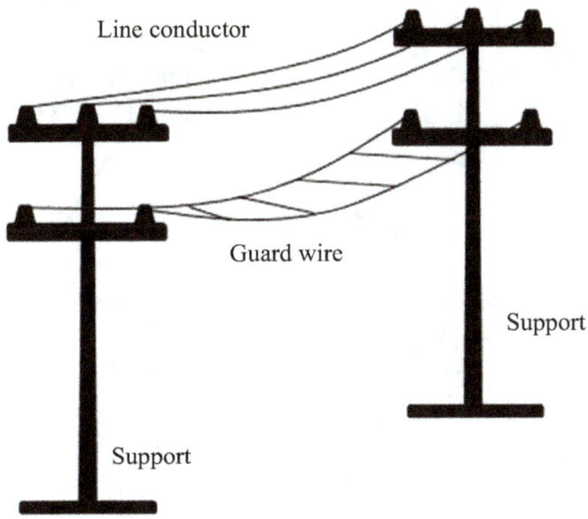

Figure 3.44 Connection of guard wire below the lines

Figure 3.45 Connection of guard wire below the 33 kV lines

3.4.12 Nuts and bolts

Nuts and bolts are small but very important units of tower used to connect different members of the tower. At present practice, they are first galvanized and then used. At the end of erection, inspection is carried out to check whether all bolts have been used and properly tightened.

3.4.13 Different shapes of tower

- I-shaped: again is of two types: narrow based and broad based (common)
- H-shaped
- V-shaped
- Inverted delta
- Y-shaped
- T-shaped

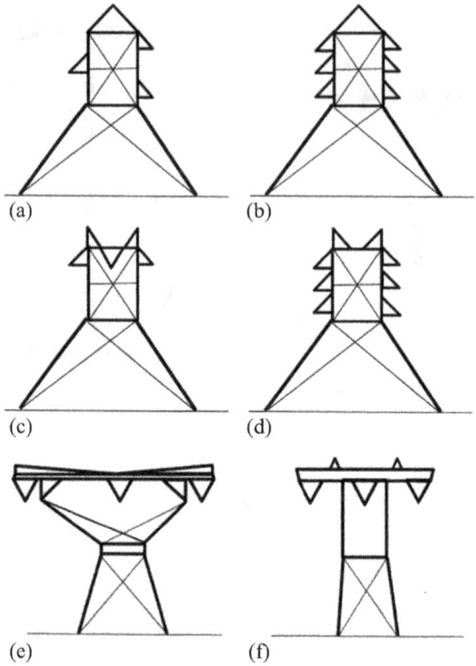

Figure 3.46 (a)–(d) Single- and double-circuit I-shaped tower, (e) wide-top tower, (f) T-shaped tower

3.4.13.1 I-shaped tower

Depending upon the base area of the tower, I-shaped towers may again be divided into two categories: narrow based and broad based. Depending on the network connection, again it is divided into two categories: single circuit and double or multiple circuit. Normally, most of the single-circuit I-shaped towers are narrow-based towers. I-shaped towers are used in LV and medium-voltage networks. Single-circuit and double-circuit I-shaped towers are shown in Figure 3.46.

Line conductors are connected through a string of overhead disc-type line insulators with the tower. The ground wire is placed at the top of the tower and is connected directly with the tower with any insulator.

3.4.13.2 H-shaped tower

Depending on the network connection, again it is divided into two categories: single circuit and multiple circuit. Normally, most of the single-circuit H-shaped towers are broad-based tower. H-shaped towers are used in high or extra-high-voltage networks. H-shaped tower is shown in Figure 3.47(a).

Normally, 20–40% of the total height remains below the ground surface level. Line conductors are connected through a string of overhead disc-type line insulators with the tower. Two ground wires are placed at the top of the tower and are connected directly with the tower without any insulator.

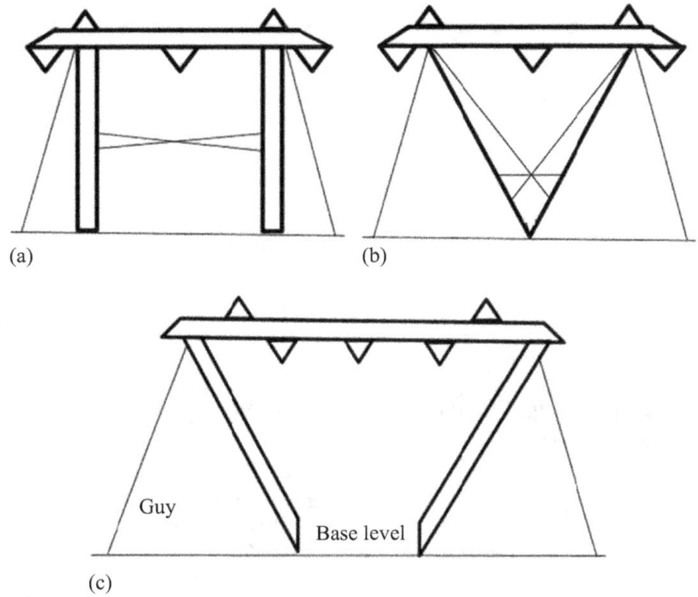

Figure 3.47 Guyed (a) H- and (b) and (c) V-shaped tower

3.4.13.3 V-shaped tower

Depending on the network connection, again it is divided into two categories: single circuit and double circuit. V-shaped towers are used in high or extra-high-voltage networks. On the base level, V-shaped towers take smaller space as shown in Figure 3.47(b). Sometimes, the base of an H-shaped tower is made narrower and given a semi-V-shaped look as shown in Figure 3.47(c).

Normally, 15–40% of the total height remains below the ground surface level depending upon the soil strength. Line conductors are connected through a string of disc-type insulators with the tower. Two ground wires are place at the top of the tower and are connected directly with the tower without any insulator (Figure 3.48).

3.4.14 Support for HVDC lines

For HVDC lines, lattice tower is used. However, a long steel pole may also be used. Narrow-based long lattice tower is suitable for HVDC lines. However, for weak soil, broad-based tower should be used. The number of cross arms for a conductor is decided by the type of HVDC line and the number of circuits carried by the transmission line. Self-supported as well as guyed tower are used. Self-supported bipolar single-circuit and bipolar double-circuit HVDC transmission towers have been shown in Figure 3.49(a) and (b), respectively. Guyed bipolar single-circuit and bipolar double-circuit HVDC transmission towers have been shown in Figure 3.49(c) and (d), respectively.

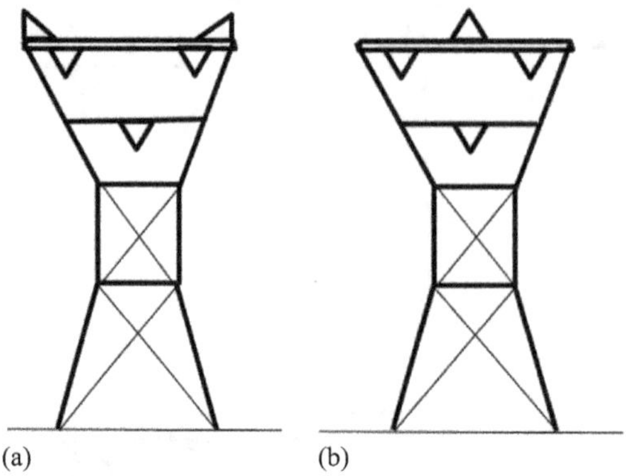

Figure 3.48 *Delta-shaped (conductor position) tower with (a) two and (b) one ground or sky wire(s)*

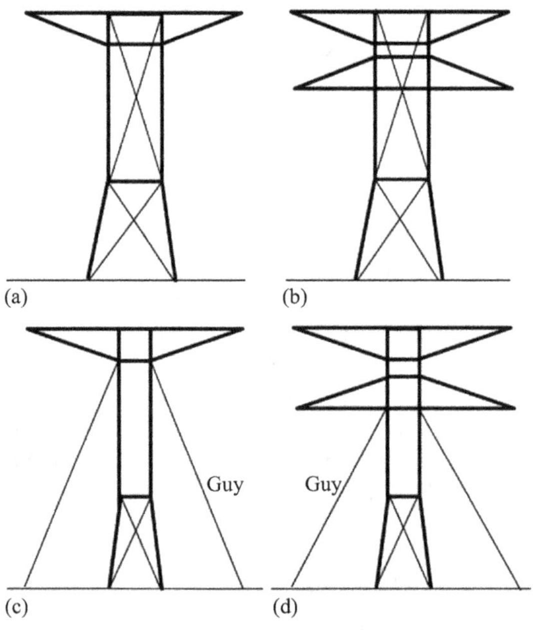

Figure 3.49 *HVDC transmission tower: (a) self-supported bipolar single circuit, (b) self-supported bipolar double circuit, (c) guyed bipolar single circuit, (d) guyed bipolar double circuit*

3.4.15 Conductor positions for tower

The following types of conductor positions are observed in a tower:

1. Vertical I for single phase
2. Vertical I-shape for three-phase four-wire system
3. Symmetrical triangular shape for three-phase three-wire system

4. Unsymmetrical triangular for three-phase three-wire system
5. Horizontal for three-phase three-wire system

Various conductor positions have been shown in Figures 3.50–3.56.

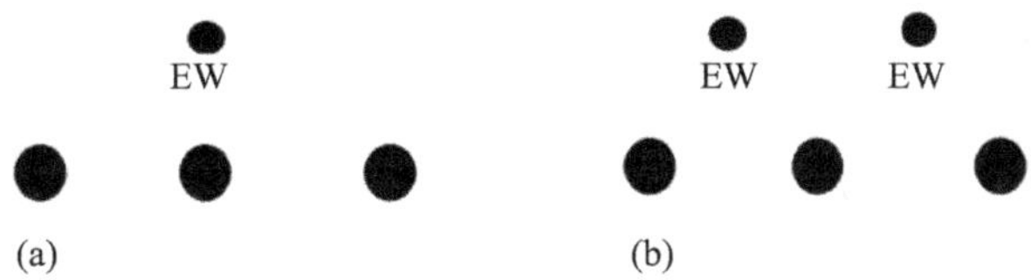

Figure 3.50 Horizontal linear conductor position with a three-wire system with (a) one earth wire, (b) two earth wires

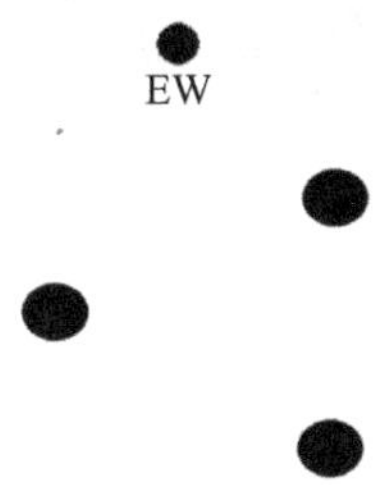

Figure 3.51 Triangular conductor position with a three-wire system with one earth wire

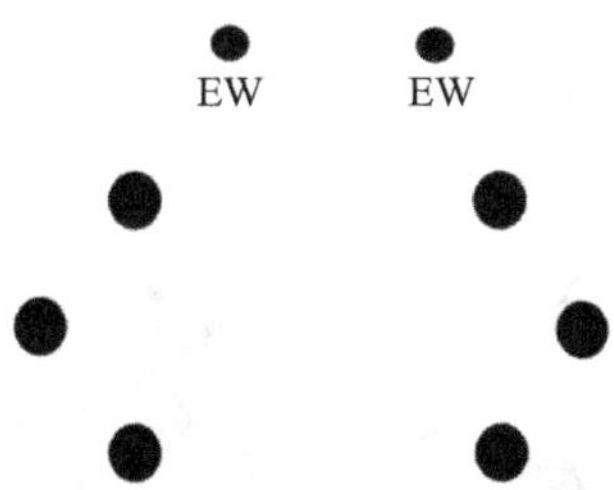

Figure 3.52 Unsymmetrical triangular conductor position with three-wire system with two earth wires for double-circuit transmission system

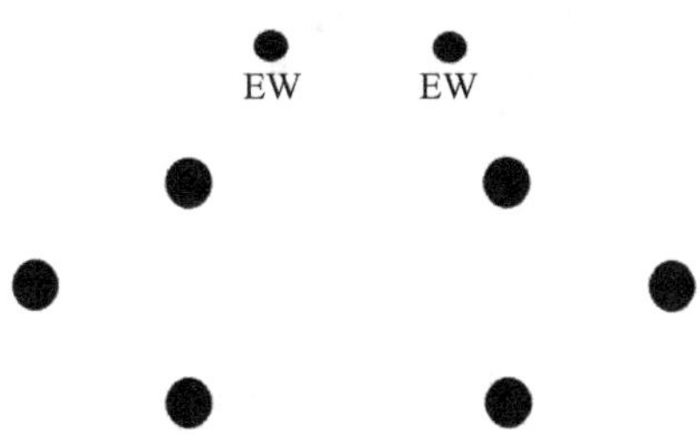

Figure 3.53 Symmetrical triangular conductor position with three-wire system with two earth wires for double-circuit transmission system

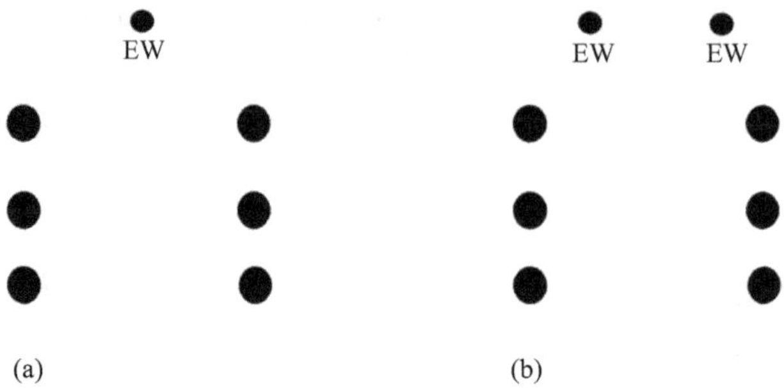

Figure 3.54 *(a) 'I'-shaped vertical linear conductor position with three-wire system with one earth wire for double-circuit transmission system; (b) 'I'-shaped vertical linear conductor position with three-wire system with two earth wires for double-circuit transmission system*

Figure 3.55 *Reverse delta triangular conductor position with three wires for double-circuit transmission system with (a) no earth wire (b) two earth wires*

Figure 3.56 *Horizontal linear conductor position with three-wire system 132 kV transmission line with two earth wires*

3.5 Foundation

3.5.1 Soil classification

Different types of soil are found in different places. Some of them are highly suitable for foundation of tower, and some are found not suitable for foundation of tower. Also soils in river side and hilly areas are different from soil of plain land.

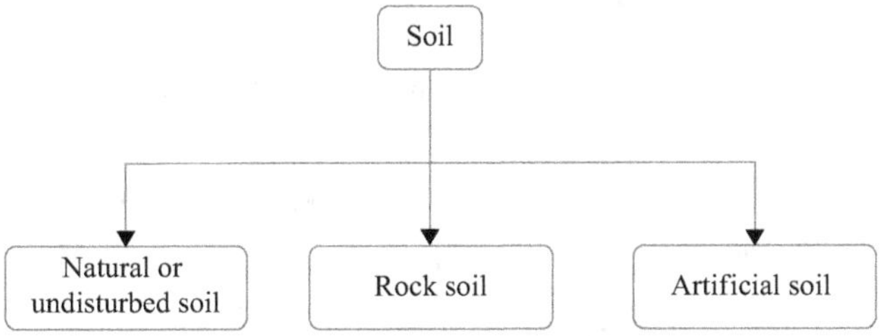

Figure 3.57 Soil classification

Therefore, before starting planning and foundation works, prior knowledge about the soil is necessary.

Overall soil classification has been shown in Figure 3.57. Soils are broadly classified into the following three categories:

1. Undisturbed soil
2. Rock soil
3. Artificial soil

3.5.1.1 Undisturbed soil

Based on chemical composition, undisturbed soil is again divided into following two main categories and shown in Figure 3.58:

1. Inorganic soil
2. Organic soil

Based on particle size, inorganic soil is again divided into three categories as follows:

1. Cohesive soil (Figure 3.59)
2. Non-cohesive soil (Figure 3.59)
3. Mixed soil

3.5.1.2 Rock soil

Rock soil refers to hard-type earth's crust particles. Rock soil is divided into the following categories:

1. Sedimentation-type rock soil – this type is originated from magma.
2. Solidification-type rock soil – this type is originated from slow sedimentation for a long time.
3. Metamorphose-type rock soil – this type is originated from thermal and mechanical deformation over a long period of time.

Different rock soils have also been shown in Figure 3.60.

Undisturbed soil

- Inorganic
  - Cohesive (particle size >0.06 mm)
  - Non-cohesive (particle size <0.06 mm)
  - Mixed
    - Cohesive dominated (non-cohesive part < 15%)
    - Non-cohesive dominated (non-cohesive part > 15%)
- Organic
  - Cohesive with greater than 5% organic parts
  - Non-cohesive with greater than 3% organic part

Figure 3.58 Undisturbed soil classification

Cohesive particles

- S: Sand particle (0.06mm – 2 mm)
  - fS: Fine sand (0.06mm – 0.2 mm)
  - mS: Medium sand (0.2mm – 0.6 mm)
  - gS: Coarse sand (0.6mm – 2 mm)
- Gravel Particle (2mm – 63mm)
  - fG: Fine gravel (2mm – 6.3mm)
  - mG: Medium gravel (6.3mm – 20mm)
  - gG: Coarse gravel (20mm – 63mm)

Non-cohesive particles

- T: Clay (< 0.002mm)
- U: Slit (0.002mm – 0.06mm)
  - fU: Fine slit (0.002mm – 0.006mm)
  - mU: Medium slit (0.006mm – 0.02mm)
  - gU: Coarse slit (0.02mm – 0.06mm)

Figure 3.59 Different cohesive and non-cohesive soil particles

Rock soil

- Sedimentation-type rock soil
- Solidification-type rock soil
- Metamorphose-type rock soil

Figure 3.60 Different rock soils

3.5.1.3 Artificial soil

Artificial soil is filled in an artificial way (planned and unplanned), most of the cases in a planned way. Soils of different nature are brought together and staged as required. In planned accumulation of soil, nature of the soil may be achieved as desired or required for particular purpose.

Suitability of commonly found soils has been presented in Table 3.3.

Table 3.3 Suitability of some commonly found soils

Soil	Property		
	Compaction	**Compressibility**	**Suitability for foundation**
Alluvial soil	Variable	Inhomogeneous	Poor–good
Boulders	Low	Unstable	Acceptable
Glacial clay	Low	High	Poor
Glacial drift (river sorted)	Average	Average-high	Good
Glacial till (unsorted)	High	Low	Good
Lateral moraine (sandy and gravelly)	Medium high	Low	Very good
Overconsolidated soil	Average	Average	Acceptable
Soft rock (weathered and un-weathered)	High	Average	Good

3.5.2 Some important terminology

3.5.2.1 Limit resistance or ultimate strength

Limit resistance or ultimate strength refers to maximum load of foundation beyond which it fails to function. The type of foundation is chosen based on limit resistance or ultimate strength of the foundation. It depends on the following features:

- Soil condition
- Type of load
- Magnitude of load
- Foundation type

Standards and guidelines are presented in EN 50341-1, 3, 4.

3.5.2.2 Characteristics foundation loads

Each foundation is characterized by load limits known as characteristics foundation loads. It is made approximately equal to 70% of designed value considered for foundation load. The design value is made less than permissible load. Permissible load is the ratio of design resistance to a partial factor of resistance.

Thus,

$$\text{Characteristic foundation load} = 0.7 \times \text{design value of foundation load}$$
$$\text{Design value of foundation load} \leq \text{permissible load}$$

$$\text{Permissible load} = \frac{\text{design resistance}}{\text{partial factor of resistance}}$$

$$\text{Partial factor of resistance} = \frac{\text{design resistance}}{\text{permissible load}}$$

3.5.2.3 Soil testing

Various methods have been introduced for soil testing. These include borings and drilling by different probes. Some commonly used probes are driven probes,

Van-type probes and compression probes. Standard penetration test is also used widely.

3.5.3 Classification of foundation for electrical support

Depending upon availability of soils and their suitability for the foundation work of electrical supports, various foundation techniques have been introduced. Foundations are broadly divided into following two categories (Figure 3.61):

1. Compact foundation
2. Separate foundation

3.5.3.1 Compact foundation

In compact foundation, all legs are established on the same compact foundation area. Monolithic foundation is the base of all types of compact foundation. This type of foundation can be used for poles (concrete, wood, steel) and also for towers. Compact foundation is mainly made by monolithic or monoblock foundation. Monolithic base caries the entire pole body or all legs of the tower. Structural load of support is transmitted into subsoil mainly by two ways:

1. Lateral earth resistance
2. Bearing pressure in foundation

Some common and popular compact foundation techniques are as follows:

1. Monoblock foundation: It is divided into two categories as follows:
   (i) Monoblock foundation without base enlargement
   (ii) Monoblock foundation with base enlargement
2. Slab foundation
3. Single-grillage foundation
4. Single-pile foundation

Different types of separate foundations have also been shown in Figure 3.62. The foundation area of a four-leg tower has been shown in Figure 3.63.

3.5.3.2 Separate foundation

In this type of foundation, for each leg of tower structure, separate foundation is made. Dead weight of the foundation and dead weight of the soil counteract uplift

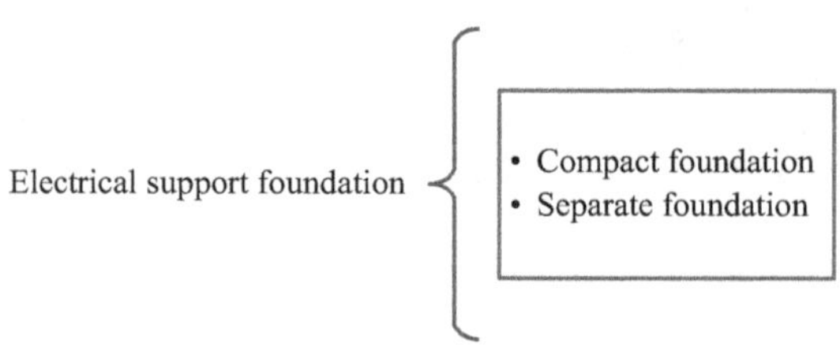

Figure 3.61 Two types of foundation used for electrical support

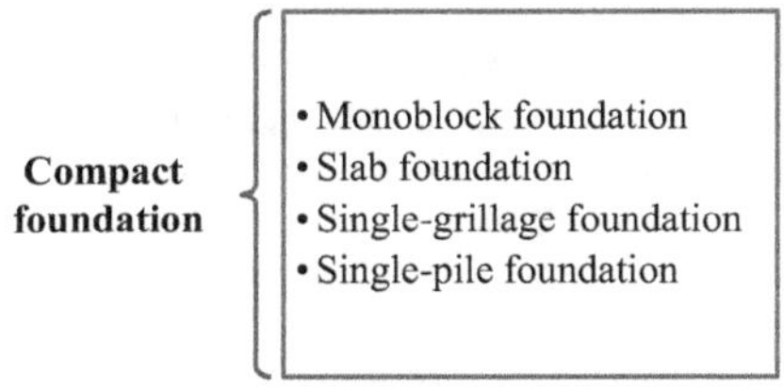

Figure 3.62 Different types of compact foundation used for electrical support

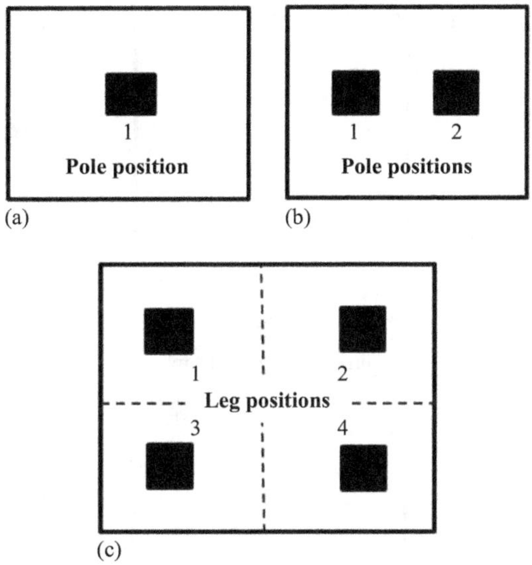

Figure 3.63 Different foundation areas for compact foundation used for electrical support: (a) foundation of single pole, (b) foundation of double pole, (c) foundation of tower with four legs

force to maintain the stability. Separate foundation is widely used for tower structure for 110 kV and above. Some common and popular separate foundation techniques are as follows:

1. Stepped block foundation
2. Auger-bored and excavated foundation
3. Separate-grillage foundation
4. Pile foundation
5. Steel reinforced pad and chimney foundation
6. Foundation in rock

Different types of separate foundations have also been shown in Figure 3.64. Foundation area of a four legs tower has been shown in Figure 3.65.

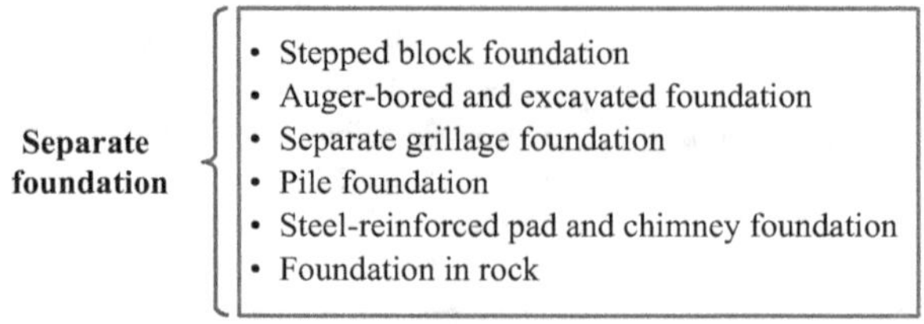

Figure 3.64 Different types of separate foundation used for electrical support

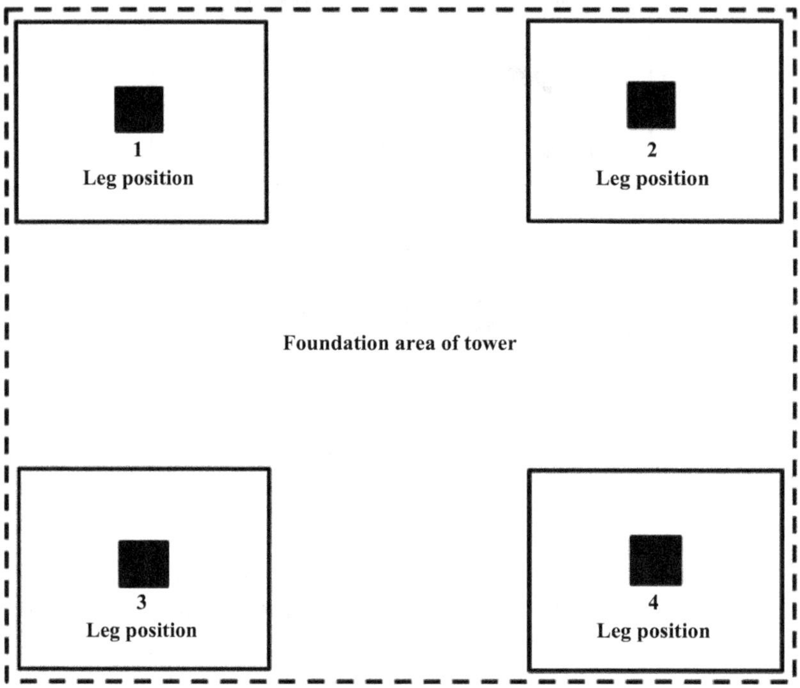

Figure 3.65 Foundation areas for compact foundation used for electric tower with four legs

3.5.4 Foundation without base enlargement

Foundation without base enlargement has been shown in Figure 3.66. Concrete foundation is made vertical, i.e. the foundation column is firmed perpendicular to foundation area at the surface or base level. The angle between column and the line connecting the bottom of column and periphery of base or surface foundation area for work is commonly known as a frustum angle. This type of foundation is used for soil of good strength and for support of less load.

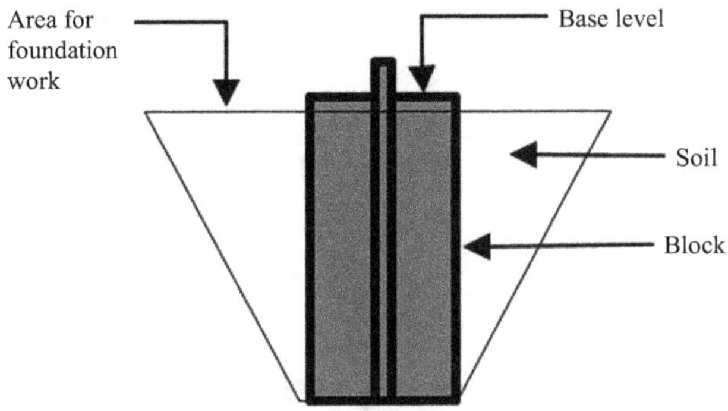

Figure 3.66 Foundation without base enlargement

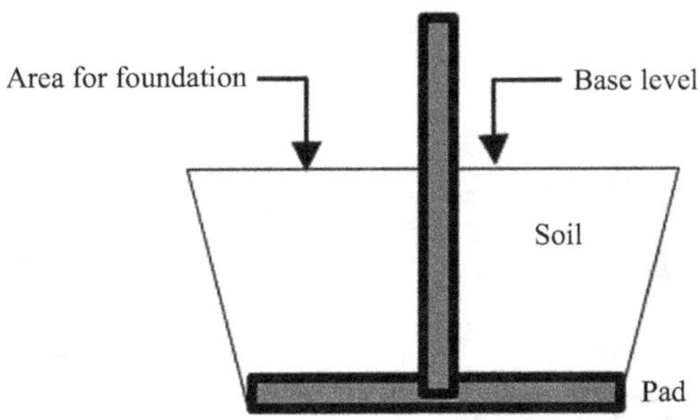

Figure 3.67 Pad foundation

3.5.5 Pad foundation

In pad foundation, concrete pad is made at bottom level and support is fixed. Pad foundation has been shown in Figure 3.67. The volume above the pad is filled tightly by soil up to base level or surface area. Pad foundation is used for support, which adds less load to foundation. Support load is distributed to the soil through pad.

3.5.6 Slab foundation

In slab foundation, a concrete slab of considerable thickness is formed above the pad at bottom level. The tower leg is inserted in it. The slab top area is filled by soil. Slab foundation has been shown in Figure 3.68. Slab foundation is found in good soil for tower of medium voltages.

3.5.7 Stepped block foundation

Stepped block foundation is widely used for soil of good strength to medium strength for medium to very large tower. At the bottom of the foundation, a pad is provided.

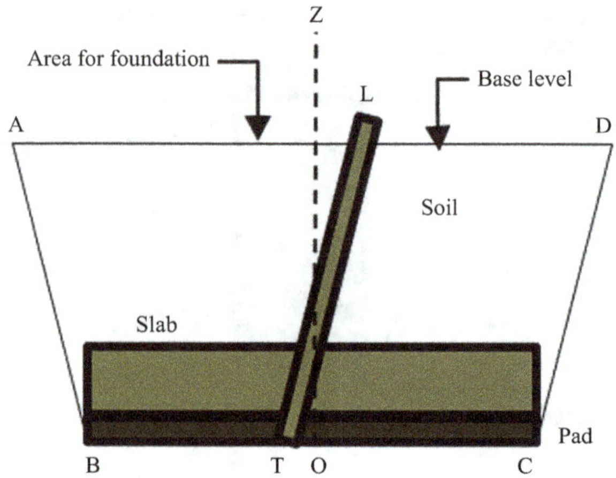

Figure 3.68 Slab foundation

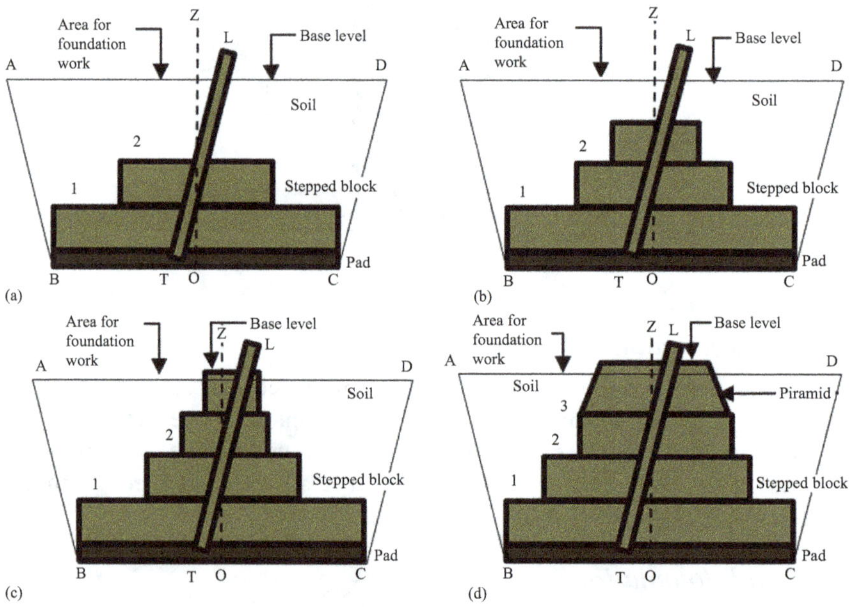

Figure 3.69 Stepped block foundation (a) two steps, (b) three steps, (c) four steps and (d) four steps with pyramid top block

Over the pad, multiple numbers of steel reinforced concrete blocks of gradually decreasing area are formed. Stepped block foundation has been shown in Figure 3.69. The top area is filled by soil. The tower leg is inserted in the blocks. Load is distributed through the blocks and bottom pad to the earth's soil. Depth and number of blocks depend on the foundation load and soils characteristics.

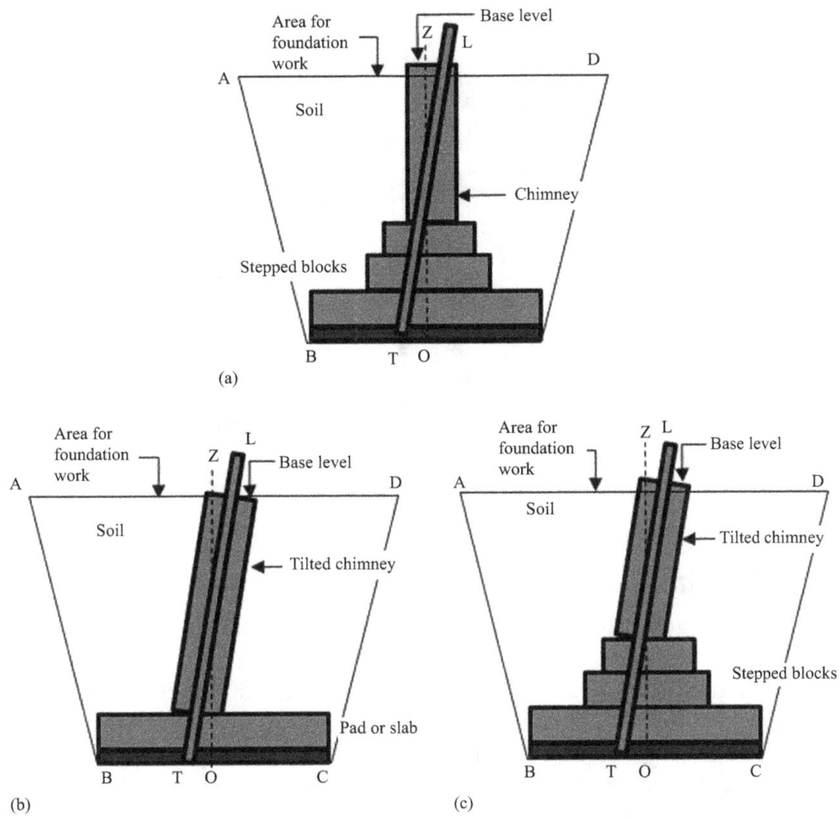

Figure 3.70 (a) Vertical chimney foundation with stepped block, (b) tilted chimney foundation with pad / slab, (c) tilted chimney foundation with stepped block

3.5.8 Pad and chimney foundation

Steel-reinforced pad and chimney foundation has become very popular choice for establishing foundation of a large tower. Long chimney shaped foundation is made over the pad. Sometimes multiple numbers of steel reinforced concrete stepped blocks are also provided and then chimney foundation is made. The leg is inserted through the chimney to blocks and pad. Load is distributed through chimney, blocks and pad to the earth. Vacant space is filled by soil. The chimney may be formed vertical or in tilted position. Different types of chimney foundations have been shown in Figure 3.70.

3.5.9 Pile foundation

Pile foundation is used for soil under river or wet area. Long depth pile is made based on the soil condition. A leg is inserted in the pile. Pile foundation has been shown in Figure 3.71.

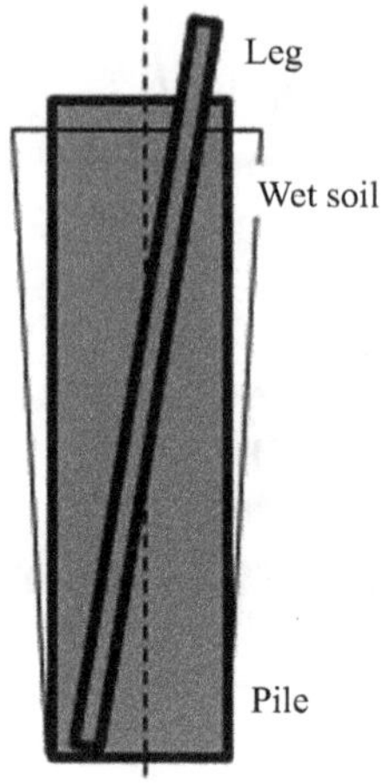

Figure 3.71 Pile foundation

3.5.10 Foundation of guyed tower and guyed wire

Foundation of a guyed tower depends on support type and soil condition. Guyed structure is important for single-pole structure or V- or H-shaped tower. Leg foundation and number of guyed wires of guyed tower are decided by tower load, tower height, conductors' load contribution, etc.

Foundation for guyed wire depends on guyed type and expected guyed load. In soil of excellent strength, and for pole support with less load, guyed wire may be inserted in the earth soil by digging up to considerable depth. However, good and safe practice is to make pad foundation or steel-reinforced column foundation, and guyed wire is inserted in it with proper slope (normally 45°). Anchor rods are used. Under the soils, anchor rods or guyed wires are covered by plastic. Embedment of guyed wire or anchor rod depends on soil type. For example, for wet soil, depth will be more than dry soil. Depth increases with the presence of more sands in soil. Depth may be less in rock soil.

3.5.11 Selection of type of foundation

The type of foundation is selected considering the following major aspects:

- Soil type
- Soil strength
- Geographical and geological nature of the region
- Foundation load
- Tower heights, conductor contribution to load, ice effect and wind effect
- Support type
- Support dimension

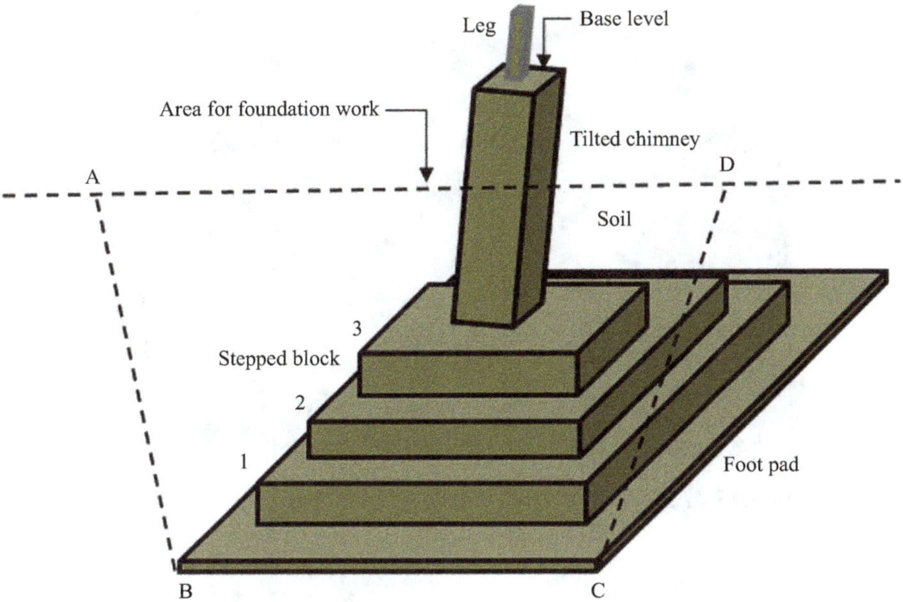

Figure 3.72 Foundation of tower of 765 kV transmission line

3.5.12 Sample foundation of a tower of 765 kV transmission line

Sample foundation of a tower of 765 kV transmission line has been shown in Figure 3.72. Chimney foundation over stepped blocks and pad has been used.

Three horizontal layers are formed below the inclined chimney column. All are steel reinforced. Steps for foundation work are as follows:

1. Identification of leg foundation area
2. Excavation of soil
3. Excavation of water if any
4. Formation of pad with steel reinforced
5. Establishment of leg member
6. Formation of blocks step by step with gradually reducing area
7. Formation of upper blocks and chimney
8. Filling of the top area by soil

A snapshot of foundation work has been shown in Figure 3.73. After completion of work, base level for the leg has been shown in Figure 3.74. Leg member coupling has been shown in Figure 3.75. Complete leg structure over leg foundation has been shown in Figure 3.76. Average depth from the ground level is 3.5 m. Steel rods (in the range of diameters of 12–35 mm) are reinforced.

At the bottom-most level, square-shaped foot pad of length 9.75 m is made. Then three gradually reducing square-shaped steeped blocks are made. A chimney column

Figure 3.73 Foundation work of a tower for 765 kV transmission line

Figure 3.74 Erection work for a leg of a tower

of 2.13 m is made above the third block. The first, second and third blocks have lengths of 6.09 m, 3.32 m, 1.82 m, respectively. Height of each block is kept 600 mm.

3.5.13 Foundation test

Different types of tests are practiced to see the foundation capacity. Broadly they may be categorized as follows:

- Design test: It refers to checking and test for the fulfilment of different design parameters and objective after foundation work.
- Proof test: It refers to test foundation capacity or range for which it is being built directly or indirectly. Test may be carried out in intermediate stage and or after completion of work.
- Parameter specific test: It is carried out to test or verify one or few specific geotechnical parameter(s) related to foundation. Main purpose of such tests is scientific research, age or future prediction, judging extensibility, etc.

Figure 3.75 Completed leg assembling of a tower

For testing purpose before use, artificial loads are imposed with the help of hydraulic system and beams of different shapes and loads.

3.6 Mechanical sag and tension

Lowering of the conductor due to its own weight is known as mechanical sag. Sag can be divided into two categories:

1. Symmetrical sag
2. Unsymmetrical sag

3.6.1 Determination of symmetrical sag

Let us consider a conductor in an x–y plane, and the centre is at the mid-point of the conductor as shown in Figure 3.76. Let tension T be divided into horizontal tension component (T_x) and vertical tension component (T_y) as follows:

$$T = T_x + T_y \tag{3.19}$$

Let, w is the weight of the conductor per unit length, S is the half-length of conductor in between two supports, L is the length between two supports, H is the horizontal pressure, h is the height of the line support.

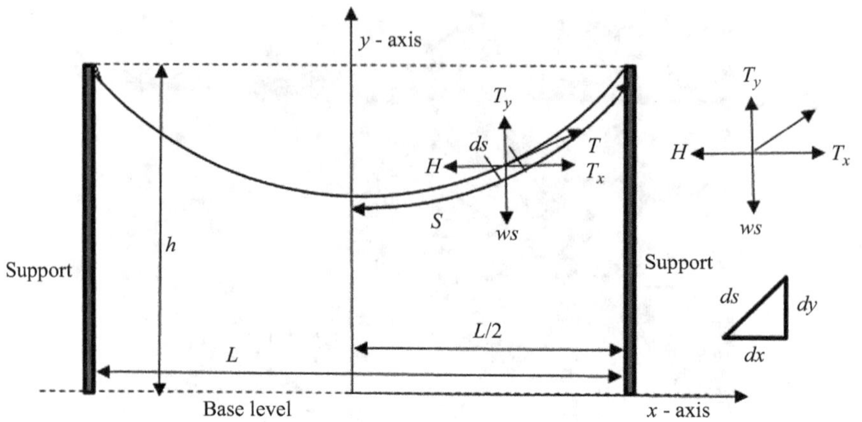

Figure 3.76 Symmetrical sag

The vertical tension is balanced by the weight and the horizontal tension is balanced by horizontal pressure. Thus,

$$T_x = H \tag{3.20}$$

$$T_y = ws \tag{3.21}$$

Therefore, slope can be expressed as follows:

$$\frac{dy}{dx} = \frac{T_y}{T_x} = \frac{ws}{H} = \sin\theta \tag{3.22}$$

Now,

$$ds^2 = dx^2 + dy^2 \tag{3.23}$$

$$dy^2 = ds^2 - dx^2 \tag{3.24}$$

$$\frac{dy^2}{dx^2} = \frac{ds^2}{dx^2} - 1 \tag{3.25}$$

$$\frac{ds^2}{dx^2} = \frac{dy^2}{dx^2} + 1 \tag{3.26}$$

$$\frac{ds}{dx} = \sqrt{\frac{dy^2}{dx^2} + 1} \tag{3.27}$$

$$\frac{ds}{dx} = \sqrt{\frac{w^2s^2}{H^2} + 1} \tag{3.28}$$

$$\frac{ds}{\sqrt{\frac{w^2s^2}{H^2}+1}} = dx \tag{3.29}$$

Solution may be written as follows:

$$x = \frac{H}{w}\sinh^{-1}\frac{ws}{H} + K \tag{3.30}$$

At $x = 0, s = 0$

$$x = \frac{H}{w}\sinh^{-1}\frac{ws}{H} \tag{3.31}$$

$$\frac{ws}{H} = \sin\frac{wx}{H} \tag{3.32}$$

$$s = \frac{H}{w}\sin\frac{wx}{H} \tag{3.33}$$

$$s = \frac{H}{w}\left\{\frac{wx}{H} + \frac{1}{6}\left(\frac{wx}{H}\right)^3 + \ldots + \ldots\right\} \tag{3.34}$$

$$\begin{aligned} s &\cong \frac{H}{w}\left\{\frac{wx}{H} + \frac{1}{6}\left(\frac{wx}{H}\right)^3\right\} \\ s &\cong \left(x + \frac{1}{6}\frac{w^2x^3}{H^2}\right) \end{aligned} \tag{3.35}$$

When, $x = L_{\text{half}} = L/2$, let $s = S$

$$S = \left(L_{\text{half}} + \frac{1}{6}\frac{w^2 {L_{\text{half}}}^3}{H^2}\right) = \left(\frac{L}{2} + \frac{1}{6}\frac{w^2L^3}{8H^2}\right) = \frac{L}{2}\left(1 + \frac{w^2L^2}{24H^2}\right) \tag{3.36}$$

Therefore, full length of the conductor may be written as follows:

$$\text{Full length} = 2S = L\left(1 + \frac{w^2L^2}{24H^2}\right) \tag{3.37}$$

Now,

$$\begin{aligned} \frac{dy}{dx} &= \frac{T_y}{T_x} = \frac{ws}{H} \\ \frac{dy}{dx} &= \frac{ws}{H} \end{aligned} \tag{3.38}$$

$$\begin{aligned} dy &= \frac{ws}{H}dx \\ dy &= \frac{w}{H}sdx \end{aligned} \tag{3.39}$$

$$dy = \frac{w}{H}\frac{H}{w} \sinh \frac{wx}{H} dx$$
$$dy = \sinh \frac{wx}{H} dx \tag{3.40}$$

Integrating,

$$y = \frac{H}{w} \cosh \frac{wx}{H} + K \tag{3.41}$$

At $x = 0, y = 0$

$$K = -\frac{H}{w} \tag{3.42}$$

Therefore,

$$y = \frac{H}{w} \cosh \frac{wx}{H} - \frac{H}{w} = \frac{H}{w}\left(\cosh \frac{wx}{H} - 1\right) \tag{3.43}$$

$$y \cong \frac{H}{w}\left\{\left(1 + \frac{1}{2}\left(\frac{wx}{H}\right)^2 + \frac{1}{8}\left(\frac{wx}{H}\right)^4 + \ldots\right) - 1\right\} \tag{3.44}$$

$$y \cong \frac{H}{w} \times \frac{1}{2}\left(\frac{wx}{H}\right)^2 = \frac{1}{2}\frac{wx^2}{H} \tag{3.45}$$

When, $x = L_{\text{half}} = L/2, y = y_{\max}$

$$y_{\max} \cong \frac{1}{2}\frac{wL_{\text{half}}{}^2}{H} \tag{3.46}$$

$$y_{\max} \cong \frac{wL^2}{8H} \tag{3.47}$$

where $y_{\max}$ is known as the sag (y_{SAG}) of the conductor at the midpoint of the span of the conductor over a length of $2L$. Therefore, mathematically sag may be written as follows:

$$y_{\text{SAG}} \cong \frac{wL^2}{8H} \tag{3.48}$$

The relation of y_{SAG} with w, L and H has been shown in Figure 3.77(a), (b) and (c), respectively.

3.6.2 Unsymmetrical sag

If the supports are placed at different base levels, maximum sag does not occur at mid of the span. The conductor is positioned in an unsymmetrical manner. This can occur in hilly regions having slope in the base levels. Unsymmetrical mechanical sag of a line conductor has been shown in Figure 3.78.

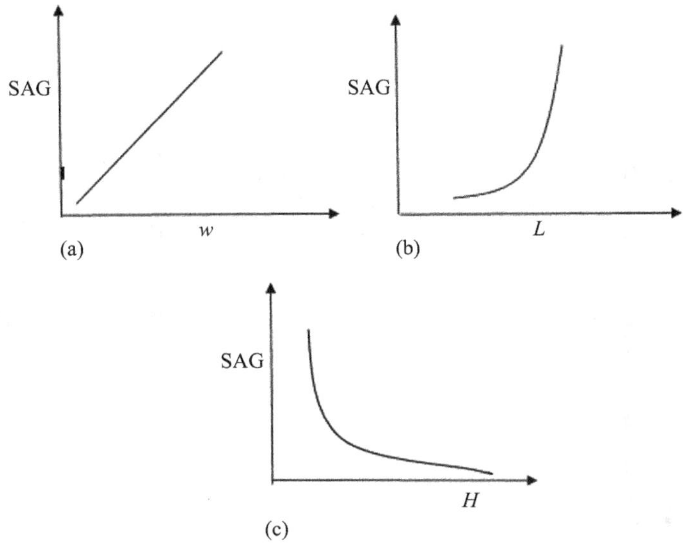

Figure 3.77 Relation of sag with (a) w, (b) L, (c) H

Unsymmetrical sag can be determined by various mathematical approaches. The maximum sag (y_{SAG_US}) can be represented as follows:

$$y_{\mathrm{SAG\_US}} = \frac{1}{2}\frac{wx^2}{T} \tag{3.49}$$

x can be determined by the following equation:

$$x = \frac{L}{2} - \frac{Ty'}{wL} \tag{3.50}$$

where y' is symmetrical sag considering the same span.

3.6.3 Clearance

If $h_{\mathrm{Cond-Top}}$ is the distance between lower conductor joint point with a line insulator to the top of the tower, ground clearance y_{GC} considering sag (y_{SAG}) and height (h) of the tower may be written as follows:

$$\text{Ground clearance} = y_{\mathrm{GC}} = h - \left(y_{\mathrm{SAG}} + h_{\mathrm{Cond-Top}}\right) \tag{3.51}$$

This ground clearance must be greater than the specified value as provided by the regulating of body of the region. It depends on the following:

- Voltage level
- Tower height
- Distance from top to lower conductor position on the tower
- Locality

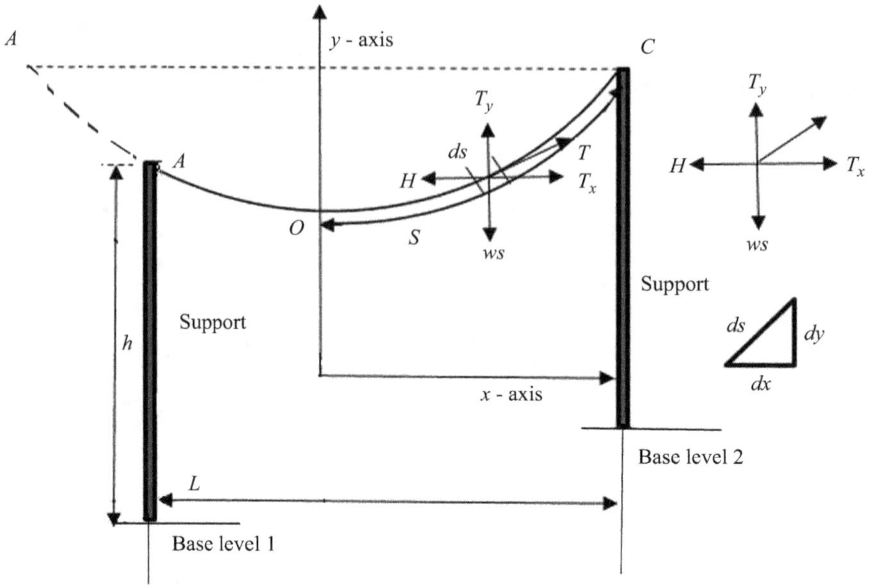

Figure 3.78 Unsymmetrical sag

- Whether road, river is there under the line
- Whether there is any construction under the line. If present, their height located
- Whether the line is crossing rail
- Whether the line is crossing over communication line

3.6.4 Effect of ice on sag

Ice fall increases the effective weight and cross-sectional area of the line conductor, which increases sag and vertical tension. In calculation of sag, weight of the ice is to be added with the weight of the conductor.

If w_c is the weight of conductor and w_i is weight of ice, then,

$$w = w_c + w_i \tag{3.52}$$

3.6.5 Effect of wind on sag

As wind flows along the horizontal direction, wind pressure does not have any effect on weight. However, force due to wind contributes to the horizontal tension H.

3.6.6 Effect of wind and ice on sag

Wind pressure increases the horizontal tension of the conductor and ice increases the effective weight of conductor. Therefore, if both wind and ice are present, then, in assessment of sag, effective wind pressure is to be added with horizontal tension of the conductor, and ice weight is to be added with the weight of conductor.

3.6.7 Sag when supports are at unequal level

When support levels are different, conductor positions of two adjusted towers differ. This makes sag of conductor unsymmetrical with respect to the span of conductor. As a result, maximum sag does not exist at the middle position of span.

Maximum sag occurs near to one of two supports.

3.7 Stringing

Installation of a conductor over towers or supports is known as stringing. It is governed by sag, tensile strength and permissible clearance from the ground and between conductors. Sag and tensile strength depend on temperature. Change of sag and tensile strength with respect to temperature variation considering the ice effect and wind effect is called or shown by a stringing chart. A typical stringing chart has been presented in Chapter 6.

3.8 Summary

Different types of lines support used in distribution and transmission system have been discussed. Installations of line poles and lattice tower have been described. Foundation methods of overhead line support have been presented. This shows two main methods: compact foundation and separate foundation. Different types of foundation categories have been described. Then, mechanical sag with tension has been presented which is needed to decide the clearance of overhead line conductors for safety.

Further reading

[1] K. O. Papailiou, Overhead Lines. CIGRE Green Books, Springer, Malters, Switzerland; 2017.

[2] E. Lakervi and E. J. Holmes, Electricity Distribution Network Design (2nd Edition). IET, Stevenage, United Kingdom; 2003, doi:10.1049/PBPO021E.

[3] Overhead Conductor Installation Guide Recommended Practices. 1st Edition, Installation guide by Electric Utility Engineering Section. General Cable Technologies Corporation, Kentucky, USA; 2014.

[4] F. Kiessling, P. Nefzger, J. F. Nolasco, and U. Kaintzyk, Overhead Power Lines. Springer, Berlin; 2003, ISBN:978-3-642-05556-0.

[5] J. Arrillaga, High Voltage Direct Current Transmission (2nd Edition). IET, Stevenage, United Kingdom; 1998, doi:10.1049/PBPO029E.

Chapter 4

Corona

This chapter deals with corona associated with high voltage (HV) or extra high voltage (EHV) overhead lines. Corona is defined, and relationship with electric field is described. Critical corona is explained. Power loss occurred due to corona in transmission system has been presented. Frequency and voltage dependencies of coronal loss have been explained. Other factors related to corona are mentioned. Practices followed for reducing corona loss have been presented.

4.1 Introduction

In the surrounding of line conductors of overhead power transmission line, there exists a gradient of electric field. To reduce the resistive loss in power transmission, voltage level is increased. But with the increase of operating voltage, field gradient also increases. For very high field gradient which is beyond the limit of withstanding capability of air, ionization takes place. This is commonly known as corona which may ultimately lead to surge or short circuit. In the process of ionization, there is a need of power that is compensated by drawing power from the transmission line. This power appears as loss of power in the transmission line and is commonly known as power loss due to corona or simply corona loss. This loss is normally undesirable in power transmission.

4.2 What is corona?

Corona of transmission system is the phenomenon of ionization of air surrounding to the conductor. Surrounding to any transmission line conductor, electric field gradient exists, and hence potential gradients as well. The gradient rises with the increase of the line potential. Due to high potential stress, ionization starts. Ionized air and free electrons move randomly and collide with air molecules and hence ionization increases. Due to the motion of ionized air molecule, 'hissing sound' is created. When ionization is very high, violet colour is observed and the phenomenon is called visual corona. It can be observed clearly at dark night at HV and EHV line. With the increase of more potential gradient, the width of ionized layer increases. Sometimes it may happen that spark or flashover is observed in between two conductors or between conductor and earth. The power required for ionization comes from transmission line and contributes to the loss of the system. This is the main disadvantage of corona.

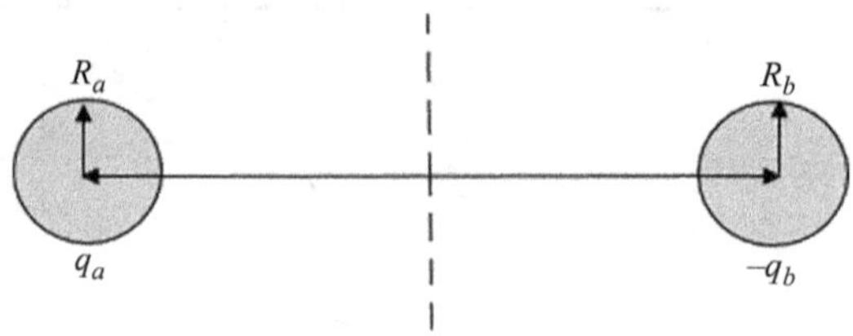

Figure 4.1 Single-phase two-wire transmission line

4.3 Voltage in a single-phase two-wire transmission line

Consider a single-phase two-wire transmission line as shown in Figure 4.1.

Let the distance between two conductors is D and their radius are R_a and R_b (located at points a and b, respectively), which is fed by single-phase AC supply which develops equal and opposite charges in two conductors, i.e.,

$$q_a = -q_b \tag{4.1}$$

Voltage will be introduced due to both charges. Voltage across a and b due to charge at a is

$$V_{ab_{q_a}} = \frac{q_a}{2\pi\varepsilon}\ln\frac{D}{R_a} \tag{4.2}$$

Similarly, voltage across a and b due to charge at b is

$$V_{ab_{q_b}} = \frac{q_b}{2\pi\varepsilon}\ln\frac{R_b}{D} \tag{4.3}$$

Therefore, net voltage across a and b can be written as follows:

$$\begin{aligned}
V_{ab} &= V_{ab_{q_a}} + V_{ab_{q_b}} \\
&= \frac{q_a}{2\pi\varepsilon}\ln\frac{D}{R_a} + \frac{q_b}{2\pi\varepsilon}\ln\frac{R_b}{D} \\
&= \frac{1}{2\pi\varepsilon}\left(q_a\ln\frac{D}{R_a} + q_b\ln\frac{R_b}{D}\right) \\
&= \frac{1}{2\pi\varepsilon}\left(q_a\ln\frac{D}{R_a} - q_a\ln\frac{R_b}{D}\right) \\
&= \frac{1}{2\pi\varepsilon}\left(q_a\ln\frac{D}{R_a} + q_a\ln\frac{D}{R_b}\right) \\
&= \frac{1}{2\pi\varepsilon}\left(q_a\ln\frac{D^2}{R_aR_b}\right) \\
&= \frac{1}{2\pi\varepsilon}\times 2q_a\ln\frac{D}{\sqrt{R_aR_b}} \\
&= \frac{1}{\pi\varepsilon}\times q_a\ln\frac{D}{\sqrt{R_aR_b}}
\end{aligned} \tag{4.4}$$

Let, $|q_a| = |-q_b| = q$ and GMR $= \sqrt{R_aR_b} = r$.

Therefore,

$$V_{ab} = \frac{q}{\pi\varepsilon} \ln \frac{D}{r} \tag{4.5}$$

Therefore, capacitance across two conductors is

$$C_{ab} = \frac{q}{V_{ab}} = \frac{\pi\varepsilon}{\ln \frac{D}{r}} \tag{4.6}$$

In between two conductors, there will be one neutral potential line and voltage across two conductors can be written as follows:

$$V_{ab} = V_{an} + V_{nb} \tag{4.7}$$

Capacitance framed with neutral plane is shown in Figure 4.2.
Net capacitance can be written as follows:

$$C_{ab} = \frac{C_{an}}{2} = \frac{C_{nb}}{2} \tag{4.8}$$

Net voltage can be written as follows:

$$\begin{aligned} V_{ab} &= V_{an} + V_{bn} = \frac{q}{\pi\varepsilon} \ln \frac{D}{r} \\ V_{an} &= V_{bn} = \frac{q}{2\pi\varepsilon} \ln \frac{D}{r} \end{aligned} \tag{4.9}$$

4.4 Electric stress in a single-phase two-wire transmission line

Electric stress refers to the voltage gradient or dielectric stress present across the surrounding dielectric medium.

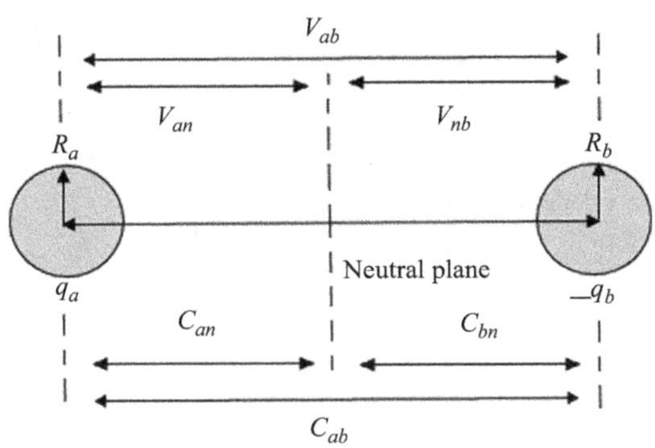

Figure 4.2 Capacitance framed by neutral plane and conductor

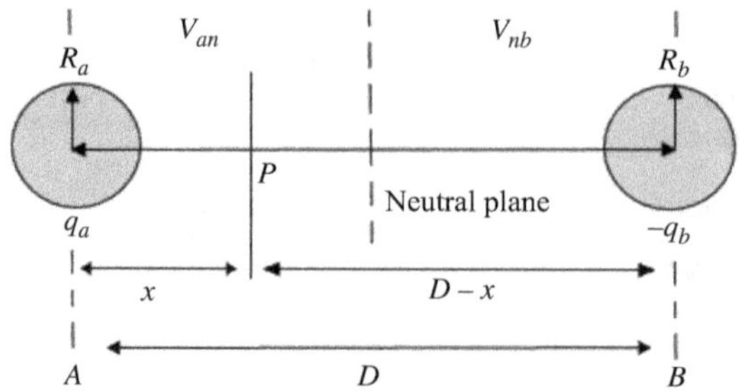

Figure 4.3 Transmission system with two conductors

Let us consider two conductors of a transmission line as shown in Figure 4.3. Distance between these conductors is D. Consider a point P in between two conductors.

Electric field at point P can be written as follows:

$$G_P = G_a + G_b \tag{4.10}$$

where G_a is the electric stress created at point P due to charge q_a and G_b is the electric stress created at point P due to charge q_b. Now,

$$G_a = \frac{1}{2\pi\varepsilon}\frac{q}{x} \tag{4.11}$$

$$G_b = \frac{1}{2\pi\varepsilon}\frac{q}{D-x} \tag{4.12}$$

The previous equations show that nearer to the conductor, electric stress will be more. Now if it is assumed that point P is located nearer to the conductor a, then D will be much in compared to x, i.e.,

$$\frac{1}{D-x} \approx 0 \tag{4.13}$$

Therefore,

$$\begin{aligned} G_P &= G_a + G_b \\ &= \frac{1}{2\pi\varepsilon}\frac{q}{x} + \frac{1}{2\pi\varepsilon}\frac{q}{D-x} \\ &= \frac{1}{2\pi\varepsilon}\left[\frac{q}{x} + \frac{q}{D-x}\right] \\ &\cong \frac{1}{2\pi\varepsilon}\frac{q}{x} \end{aligned} \tag{4.14}$$

Let, $r = \sqrt{R_a R_b}$.

This electric stress will be maximum near the surface of the conductor, i.e., at $x = r$. Then,

$$\begin{aligned} G_{\max} &= G_a + G_b \\ &= \frac{1}{2\pi\varepsilon}\frac{q}{x} + \frac{1}{2\pi\varepsilon}\frac{q}{D-x} \\ &= \frac{1}{2\pi\varepsilon}\left[\frac{q}{x} + \frac{q}{D-x}\right] \\ &\cong \frac{1}{2\pi\varepsilon}\frac{q}{x} \\ &= \frac{1}{2\pi\varepsilon}\frac{q}{r} \end{aligned} \tag{4.15}$$

From voltage equation we can write

$$V_{ab} = V_{an} + V_{bn} = \frac{q}{\pi\varepsilon}\ln\frac{D}{r} \tag{4.16}$$

Therefore,

$$G_{\max} = \frac{1}{2\pi\varepsilon}\frac{q}{r} = \frac{V_{ab}}{2r\ln\frac{D}{r}} \tag{4.17}$$

Also,

$$V_{an} = V_{bn} = \frac{q}{2\pi\varepsilon}\ln\frac{D}{r} \tag{4.18}$$

Therefore,

$$G_{\max} = \frac{1}{2\pi\varepsilon}\frac{q}{r} = \frac{V_{an}}{r\ln\frac{D}{r}} = \frac{V_{bn}}{r\ln\frac{D}{r}} \tag{4.19}$$

4.4.1 Corona voltage

Phase to neutral voltage may be expressed in terms of maximum electric stress as follows:

$$V = V_{bn} = V_{bn} = G_{\max} r \ln\frac{D}{r} \tag{4.20}$$

Voltage at which ionization starts is known as corona voltage (V_c) which can be expressed as

$$V_c = G_0 r \ln\frac{D}{R} \tag{4.21}$$

where, G_0 is the electric stress when corona starts.

At air pressure of 76-cm mercury and 23°C temperature, electric stress or voltage gradient required to ionize air or break the air is

$$G_0 = 30\ \text{kV/cm(max) or } 30\ \text{kV/cm (r.m.s.)}$$

Electric stress depends on the density of the air and this relation is proportional. Mathematically stress can be written as follows:

$$G = G_0\delta \tag{4.22}$$

where δ is known as air density factor. If b be the barometric air pressure and t be the temperature, then air density factor can be written as follows:

$$\text{air density factor } \delta = \frac{3.92\, b}{273 + t} \tag{4.23}$$

when b is 76-cm and t is 23°C, $\delta = 1$.

Therefore, considering air density factor,

$$V_c = Gr\ln\frac{D}{R} = G_0\delta r\ln\frac{D}{R} = G_0\frac{3.92\, b}{273 + t}r\ln\frac{D}{R} \tag{4.24}$$

Further, electric stress depends on the nature of the conductor surface. Considering this, an irregularity factor is introduced in the expression as follows:

$$G = m_0 G_0\delta \tag{4.25}$$

where, for polished surface, $m_0 = 1$; for dirty surface, $m_0 = 0.92$ to 0.98; for surface of stranded conductor, $m_0 = 0.8$ to 0.87.

Therefore, corona voltage can be written as follows:

$$V_c = m_0 G_0\delta r\ln\frac{D}{R} = m_0 G_0\frac{3.92\, b}{273 + t}r\ln\frac{D}{R} \tag{4.26}$$

Above this voltage corona occurs. Often this voltage is referred to as critical corona voltage or disruptive corona voltage.

Minimum phase to neutral voltage at which visual effect or glow is observed due to ionization is known as visual corona voltage. Mathematically, visual corona voltage is expressed as follows:

$$V_v = m_0 G_0\delta\left(1 + \frac{0.3}{\sqrt{\delta r}}\right)\ln\frac{D}{R} \tag{4.27}$$

4.5 Power loss

Major disadvantage of corona is power loss because power required for ionization of surrounding air and discharge is drawn from the system. This power loss is known as power loss or simply corona loss. Two empirical formulas are used to determine this loss: Peak's formula and Peterson's formula. Peak's formula is used when ratio V/V_C is less than 1.8; otherwise Peterson's formula is more useful.

Peak's formula may be written as follows:

$$Power\ Loss_{corona} = P_C = 21 \times \frac{1}{\left\{\ln\frac{D}{r}\right\}^2} f V^2 F \times 10^{-6}\ \text{kW} \tag{4.28}$$

Peterson's formula may be written as follows:

$$Power\ Loss_{corona} = P_C = \frac{244}{\delta}(f + 23)\sqrt{\frac{r}{D}}(V - V_C)^2 \times 10^{-3}\ \text{kW} \tag{4.29}$$

Factor F is constant and it depends on the ratio V/V_C.

Equations (4.28) and (4.29) show that corona loss increases with the square of the voltage indicating that it will be high in EHVAC lines. Also it shows that corona has two parts: (i) frequency-independent part indicating that corona loss will be present in DC line also with comparatively less magnitude and (ii) frequency-dependent part indicating that corona loss in AC is more than that of DC lines.

4.6 Factors of corona

4.6.1 Frequency

Power loss due to corona has two parts: one depends on frequency and the other is independent of frequency. Due to the frequency-dependent part corona and corona loss increases with the increase of frequency. When frequency becomes zero, loss becomes minimum. Thus corona in AC is greater than corona in DC. That is why EHVAC transmission system is being converted into HVDC transmission system.

4.6.2 Voltage

With the increase of voltage, corona increases. Corona loss is proportional to the square of the voltage as shown in Figure 4.4. Therefore, corona loss is more in EHVAC lines than that of HVAC lines.

4.6.3 Dust

Overhead transmission lines are exposed in air and dust particles may get accumulated on the conductor surface. Dust on the conductor surface reduces critical voltage. Thus, with the increase of dust on the surface of conductor, corona loss increases.

4.6.4 Rain

Rain reduces the conductivity of air and reduces dielectric strength. It helps in ionization process. Thus rain increases corona loss. However, at heavy rainfall moisture contents get saturated and corona loss remains constant and does not

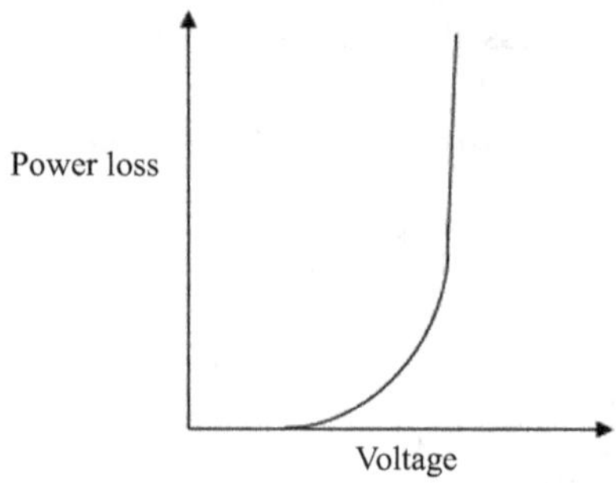

Figure 4.4 Voltage versus power loss due to corona

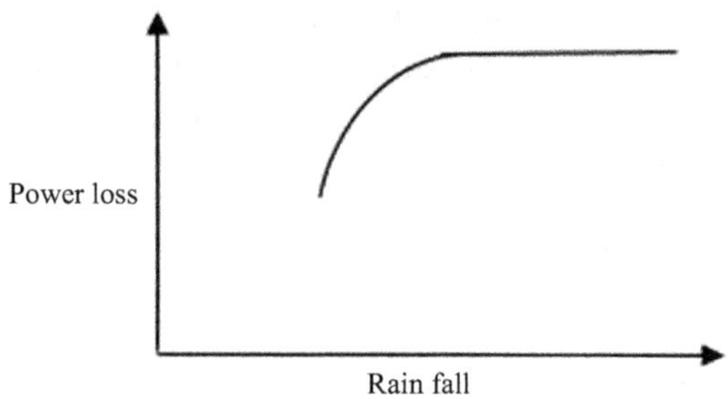

Figure 4.5 Rain fall versus power loss due to corona

increase any more. A typical relation between rainfall and corona loss is shown in Figure 4.5.

4.6.5 Snow or hail effect

Snow or hail effect reduces critical voltage. Thus, with the increase of snow or hail effect, corona loss increases.

4.6.6 Atmospheric temperature

Overhead transmission lines are exposed in atmosphere. The rise of atmospheric temperature reduces moisture content on the surface of conductor. Thus, it decreases corona loss.

4.6.7 Load

With the increase of load, copper loss increases. It increases the temperature on the surface of the conductor. This reduces moisture content of the air and reduces ionization. Thus, with the increase of load, corona loss increases.

4.7 Methods of reducing corona

1. Distance: With the increase of distance between two conductors, corona reduces. Thus at the time of design, distance is to be kept as maximum as possible to reduce corona loss.
2. Radius of conductor: With the increase of radius of line conductor, corona increases. Thus at the time of design, radius of line conductor is to be kept as short as possible to reduce corona loss.
3. Voltage: With the increase of line voltage, corona reduces. Thus at the time of design, line voltage is to be kept as low as possible to reduce corona loss.
4. HVDC: Frequency-dependent term is absent in the mathematical expression of corona loss of DC line. Thus corona loss becomes very small in HVDC line. So, EHVAC line can be replaced by HVDC line to reduce corona loss of the system.
5. Bundled conductor: Use of bundled conductor reduces average corona loss both in HVAC and HVDC lines.
6. Corona ring: Application of corona rings in overhead lines reduces corona loss.

4.8 Corona ring

Corona ring is widely used in overhead transmission lines with transmission voltage 220 kV or above to reduce corona discharge. It is made of conducting materials of ring shape of a larger diameter. It is electrically connected with line conductor and is kept at conductor potential. It is positioned at conductor end of string. Corona discharge depends on gradients of electric field which is high near the surface of conductor. If there is any sharp end or sharp surface area, electric field gradient increases. If the potential gradient exceeds dielectric strength of air or critical disruptive voltage of air, plasma of ionized air forms and corona discharge takes place. It appears as power loss to the power system. Corona ring reduces loss by reducing potential or electric field gradient near the surface area.

4.9 Disadvantages

1. *Loss:* Power required for ionization of air comes from the transmission line. Thus it is treated as loss of the system and known as corona loss. At HV or EHV, corona loss is very high. This is the main disadvantage of corona.
2. *Capacitive current:* Due to corona, line capacitance increases. This increases charging and discharging current of the line.
3. *Triple-frequency component:* Triple-frequency current flows in grounded AC system. It also generates triple frequency in voltage waveforms.

4.10 Advantages of corona

Corona in most of the cases is highly undesirable as it increases loss. However, it has few advantages. High voltages are generated by lightening effect or switching effect. They disappear through the ionization process of corona. Thus it helps in providing protection against HV due to switching, lightening, etc. In some manufacturing process, corona discharge is utilized.

4.11 Corona in HVDC lines

As in HDVC transmission, frequency of the voltage becomes zero, corona loss is much less than HVAC lines. However, amount of corona loss in HVDC is not zero. Moreover, corona loss is found to be increasing with the increase of voltage. The loss decreases with the increase of pole spacing of the line. Use of bundled conductor reduces average corona loss; however, maximum corona loss is almost same for bundled (double) and single conductors.

4.12 Research advancement

Corona on overhead lines has become an interest of study to the power researchers. Different test methods have been introduced for AC as well as HVDC [1]. Discharge mainly takes place over surface in overhead lines [2,3]. Different computational techniques have been introduced for the determination of power loss due to corona discharge [4]. Current nature [5] during corona discharge in HVAC or HVDC lines has been modelled using EMTP software [6,7]. Corona discharges related to other surges in transmission lines have been modelling with charging theory [8]. Corona with icing effects [9], bundled conductor [10] and lightning [11] have become areas of research interest. Thus corona in overhead lines gives ample scope for study to evolve better tests and measurement methodology, to reduce undesirable losses and damages and to apply corona phenomenon in protection and other need-based applications.

4.13 Summary

This chapter presents the phenomenon of corona. It has explained how corona appears in overhead lines. Voltage dependency has been presented. Loss due to corona increases with the square of the voltages. Therefore, in HVAC and EHVAC lines, loss due to corona shares major part of total loss of the transmission system. Other factors influencing the corona loss have been presented. It shows that rain, snow, fog, etc. increase corona loss significantly. Different ways of reducing corona loss have been presented. Different features of corona in HVDC lines have been presented.

4.14 Standard

Some useful standards on corona discharge have been presented in Table 4.1.

Table 4.1 Some useful standards on corona discharge

IEEE standard	Purpose
IEEE Std 436-1977	IEEE guide for making corona (partial discharge) measurements on electronics transformers
IEEE Std 454-1973	IEEE recommended practice for the detection and measurement of partial discharges (corona) during dielectric tests
IEEE Std C37.301-2009	IEEE standard for high-voltage switchgear (above 1,000 V) test techniques – partial discharge measurements
IEEE Std 430-2017	IEEE standard procedures for the measurement of radio noise from overhead power lines and substations

References

[1] B. M. Bailey, 'Test line experience with HVDC overhead transmission'. in IEEE Transactions on Power Apparatus and Systems, vol. PAS-89, no. 7, pp. 1625–1634, 1970.

[2] T. Klueter, J. Wulff, F. Jenau, and D. Wienold, 'Evaluation of surface- and corona discharges at DC voltage'. 2013 13th International Conference on Environment and Electrical Engineering (EEEIC), Wroclaw, 2013, pp. 255–259, doi: 10.1109/EEEIC-2.2013.6737918.

[3] Y. Yi, Y. Wang, and L. Wang, 'Conductor surface conditions effects on audible noise spectrum characteristics of positive corona discharge'. in IEEE Transactions on Dielectrics and Electrical Insulation, vol. 23, no. 3, pp. 1872–1878, 2016, doi: 10.1109/TDEI.2016.005399.

[4] X. Li, O. P. Malik, and Z. Zhao, 'Computation of transmission line transients including corona effects'. in IEEE Transactions on Power Delivery, vol. 4, no. 3, pp. 1816–1822, 1989, doi: 10.1109/61.32677.

[5] M. Pfeiffer, S. Hedtke, and C. M. Franck, 'Corona current coupling in bipolar HVDC and hybrid HVAC/HVDC overhead lines'. in IEEE Transactions on Power Delivery, vol. 33, no. 1, pp. 393–402, 2018, doi: 10.1109/TPWRD.2017.2713603.

[6] T. J. Gallagher and I. M. Dudurych, 'Model of corona for an EMTP study of surge propagation along HV transmission lines'. in IEE Proceedings – Generation, Transmission and Distribution, vol. 151, no. 1, pp. 61–66, 2004, doi: 10.1049/ip-gtd:20030927.

[7] U. Corbellini and P. Pelacchi, 'Corona losses in HVDC bipolar lines'. in IEEE Transactions on Power Delivery, vol. 11, no. 3, pp. 1475–1481, 1996, doi: 10.1109/61.517506.

[8] M. Afghahi and R. J. Harrington, 'Charge model for studying corona during surges on overhead transmission lines'. in IEE Proceedings C – Generation, Transmission and Distribution, vol. 130, no. 1, pp. 16–21, 1983, doi: 10.1049/ip-c.1983.0003.

[9] G. He, Q. Hu, L. Shu *et al.*, 'Impact of icing severity on corona performance of glaze ice-covered conductor'. in IEEE Transactions on Dielectrics and Electrical Insulation, vol. 24, no. 5, pp. 2952–2959, 2017, doi: 10.1109/TDEI.2017.006536.

[10] Z. Al-Hamouz and M. Abdel-Salam, 'Finite-element solution of monopolar corona on bundle conductors'. in IEEE Transactions on Industry Applications, vol. 35, no. 2, pp. 380–386, 1999, doi:10.1109/28.753632.

[11] C. A. Nucci, S. Guerrieri, M. T. Correia de Barros, and F. Rachidi, 'Influence of corona on the voltages induced by nearby lightning on overhead distribution lines'. in IEEE Transactions on Power Delivery, vol. 15, no. 4, pp. 1265–1273, 2000, doi: 10.1109/61.891513.

Chapter 5
Overhead line insulator

This chapter describes different aspects of overhead insulators. Properties of line insulators have been described. Materials used for manufacturing overhead line (OHL) insulators have been presented. Different types of insulators used in OHL application have been discussed. Requirement of string has been mentioned. Voltage distribution across different discs of string has been shown, and string efficiency has been derived. Design features of insulators for overhead application have been presented. Selection factors for insulator have been mentioned for transmission and distribution line applications. Fittings of insulator in OHLs have been described. Failure and testing of OHL insulator have been discussed. Different useful standards for this aspect have been referred.

5.1 Introduction

Overhead power transmission line has three major parts: line support, conductor and line insulator. Support is a mechanical structure used to hold the conductors with certain clearance from the ground. Conductor is the main part of overhead line (OHL) used to carry electric power. Line insulator is the remaining important part of overhead transmission line located in between line support and conductor. All the three parts are essential and should be in healthy condition for power transmission. Type and number of line insulators depend on operating voltage of the corresponding transmission system.

5.2 Overhead line insulator

Overhead line (OHL) insulator provides insulation between high-voltage bare conductor and the near-earth point of support. It also provides mechanical support to the bare conductors.

5.3 Common properties of line insulator

1. High resistivity: OHL insulator should have high resistivity.
2. Dielectric strength: OHL insulator should have high dielectric strength which increases the voltage withstanding capability.

3. Mechanical strength: As OHL insulator also provides support to the bare conductor, it should have the capacity of withstanding the tension of carrying bare conductors. For this purpose, line insulators should have high mechanical strength.
4. Leakage resistance: Leakage current may flow over the surface which is not desirable. Good line insulator should not have leakage current over it. Thus, it should have very high leakage resistance. For this purpose, cleavages are provided to increase leakage path and hence leakage resistance. Higher the number of cleavages, higher the leakage resistance. Surface is made highly polished to increase the leakage resistance.
5. Porosity: Porosity is not desirable in OHL insulator.

Properties of OHL insulators have been discussed in detail in the following sections.

5.4 Material of overhead line insulators

Materials used for OHL insulators can be categorized in the following three groups:

1. Porcelain
2. Glass
3. Composite silicone

5.4.1 Porcelain

Use of porcelain as line insulator has started since 1850. Porcelain material is available plenty in nature. It shows high dielectric strength, electrical and thermal resistivity which has made it a good choice as insulator. Low surface leakage and porosity are obtained by proper design. It also offers good mechanical strength. Examples of some OHL insulator made of porcelain have been shown in Figure 5.1.

Figure 5.1 Porcelain insulators

Figure 5.2 Glass insulators

5.4.2 Glass

Glass material also shows high dielectric strength, electrical and thermal resistivity which has made it another good choice as insulator. Use of glass as line insulator has started since the eighteenth century. Glass is available plenty in nature. The advantage of glass material is that they are transparent which makes the visual inspection of the material quality with respect to crack, air pockets, etc. much easier than porcelain material which is not transparent. Low surface leakage and porosity are obtained by proper design. It also offers good mechanical strength. OHL insulator made of glass have been shown in Figure 5.2.

5.4.3 Composite silicone

With the development of material science, silicone and their composites have been introduced in electronic applications as well as in electrical high-voltage applications. Different composites have been found suitable for underground cable application and OHLs insulation. Composite silicone insulation has become an alternate in some cases preferred choice replacing the use of porcelain and glass for the purpose. Composite silicone insulators [1] find applications in low-voltage as well as high-voltage application. OHL insulator made for low voltage has been shown in Figure 5.3. It can be customized as required for particular OHLs.

5.5 Classification of overhead line insulators

Depending on the construction and use, OHL insulators are divided into the following two categories:

1. Pin type
2. Disc type

Figure 5.3 Composite material for low voltage application

Figure 5.4 Pin-type insulator

5.5.1 Pin-type insulator

Pin-type insulator is shown in Figure 5.4. Conductor is placed at the top of the insulator. The surface is highly polished so that less amount of dust will be accumulated on the surface which will absorb less moisture. Also polished surface will drop raindrops quickly. The shape consists of many cleavages. The presence of cleavages in the shape of the insulator increases effective length of cleavage path. Hence, it increases leakage resistance and decreases leakage current. So, if the number of cleavages increases, cleavage resistance will increase which will decrease leakage current. Pin-type insulators are mainly used in low-voltage and medium-voltage range. Application of pin-type insulator is more at below 33 kV. However, at substation, pin insulator is often used at higher voltage level.

Pin-type insulator made of porcelain connected with 33 kV and 11 kV line has been shown in Figure 5.5(a) and (b), respectively. Lower end is connected with cross arm which is held tightly with the pole by clamp arrangement. Conductor is placed at the top of the insulator. Pin-type insulator made of ceramic material connected with 440 V distribution line has been shown in Figure 5.6. Lower end is connected with the pole by clamp arrangement. Conductor is placed at the top of the insulator.

5.5.2 Disc-type insulator

Disc-type insulator is shown in Figure 5.7. Conductor is placed at the bottom of the insulator. The surface is highly polished so that less amount of dust will be

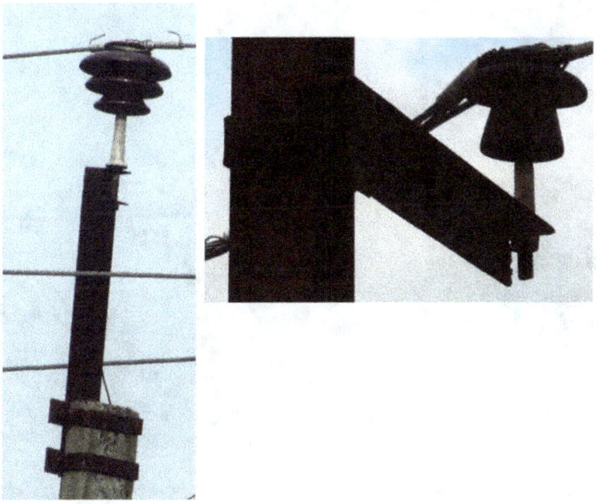

(a) (b)

Figure 5.5 Pin-type insulator connected with (a) 33 kV line and (b) 11 kV line

Figure 5.6 Pin-type insulator connected with 440 V line

Figure 5.7 Disc-type insulator

Figure 5.8 Disc-type insulator strained in 11 kV line

accumulated on the surface, which will absorb less moisture. Also polished surface will drop raindrops quickly. The shape consists of many cleavages. The presence of cleavages in the shape of the insulator increases effective length of leakage path. Hence, it increases leakage resistance and decreases leakage current. So, if the number of cleavages increases, leakage resistance will increase which will decrease leakage current. Dingle disc insulator is mainly used in low-voltage and low-medium-voltage range. Operating voltage of single disc insulator is normally 11 kV. However, it may be higher for specially designed discs. In practice, more than one disc is used together at medium, high, extra high-voltage (EHV) levels.

Disc-type insulator is connected either in suspension or in strain condition. Strain-type disc insulator connected with 11 kV distribution line has been shown in Figure 5.8. Disc is held tightly with pole by clamp arrangement at one end and at other end, it is connected with conductors of line and jumper. Discs are found in different forms:

- Interlinking type (also known as Hewlett type)
- Cemented cap type
- Cap and pin type
- Core and link type

5.5.3 Shackle-type insulator

Shackle-type line insulators are used in low-voltage distribution lines. They are held by the D- or U-shaped clamp arrangement from support. Conductor is placed in the middle position and bounded. One or multiple numbers of shackle-type insulators are used as required by the application. Shackle-type insulator has been shown in Figure 5.9.

Figure 5.9 Shackle-type insulator strained in 11 kV line

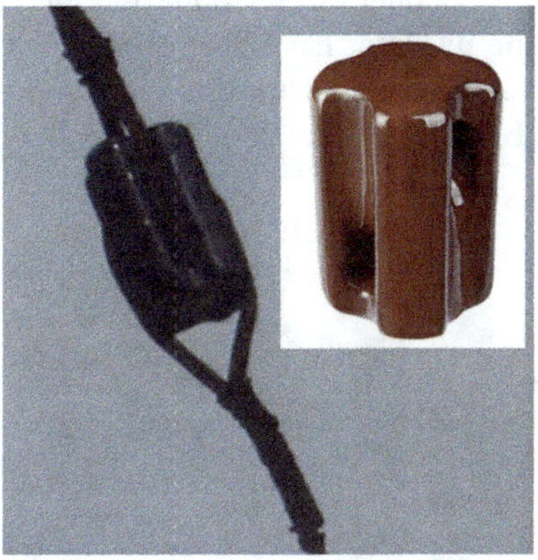

Figure 5.10 Stay-type insulator

5.5.4 Stay-type insulator

Stay-type line insulators are used in guy. They are used to join two parts of the guy wire above the ground. Stay-type insulator has been shown in Figure 5.10. Voltage withstanding capability is comparatively lower than other types of line insulators.

5.5.5 Line-post-type insulator

Line-post insulators are used at the end or starting of transmission line in substation. Length of single unit of line-post insulator is much higher than the others. They can be placed on self-supporting manner.

5.5.6 Porcelain long-rod-type insulator

Porcelain long-rod-type insulators are found for very low maintenance requirement. Their weight is lighter than cap- and pin-type insulators. They are made as puncture proof and show high electrical and mechanical stress withstanding properties. However, their production cost is high and shows lower strength against impulsive mechanical force. Their applications are mostly found in Europe and Middle East.

5.5.7 Composite silicone insulator for transmission line

A typical composite overhead silicone insulator has been shown in Figure 5.11(a). Along the central axis, there is glass fibre rod offering insulation and high mechanical strength in axial direction. It is surrounded by housing made of silicone rubber which offers high resistance to different climatic conditions as follows:

- It is highly resistant to water; water cannot penetrate.
- It is inflammable.
- It is chemically inert.
- It offers high heat withstanding property.
- It is less affected by UV rays and sunlight.

At the two ends, metal conductors are connected with different shapes. Socket and ball ends of composite insulator have been shown in Figure 5.11(b). They are joined by lock, as shown in Figure 5.11(c). Ends of composite silicone insulator as string for application in EHV transmission have been shown in Figure 5.11(d).

Figure 5.11 Parallel string in 765 kV line: (a) single string, (b) two ends of string, (c) jointing of two string with lock, (d) four parallel combination of two cascaded string connected with cross arm and bundled conductor of 765 kV transmission line

Table 5.1 Comparison of different types of overhead line insulators

Type	Pin type	Cup and pin (disc)	Porcelain long rod	Composite silicone long rod
Mechanical strength	Low	High	Low	High in axial direction
Electrical stress		Low	High	High
Design flexibility	Low	Flexible	Not flexible	Customized
Weight	Low	High	High	Low
Puncture risk	Low	High	Low	Low
Corrosion susceptibility	Low	High	Low	Medium
Hydrophobia	Low	Low	Low	High
Production cost	Low	Low	High	High
Application	Low-to-medium-high distribution	High to extra high	High to extra high	Medium to ultra-high

5.5.8 Comparison of different types of insulators

Comparison of different types of OHL insulators has been presented in Table 5.1.

5.6 Requirement of insulator sets

Limitation of using single insulator is that if the insulator gets damaged by any reason, to run the system, immediate replacement of the insulator is to be carried out. Sometimes, it becomes difficult to implement it in a time-bound manner. But, instead of single insulator, set(s) of insulators is used. If one weak most single insulator gets damaged, even then the system can be run over a period of time; sufficient time will be in hand to replace the damaged one. For this purpose, disc insulators are chosen to make required set of combination of insulators.

5.7 String of insulators

Sets or combination of disc insulators connected one after another are known as string or string of insulators.

Disc-type insulators are divided into the following two categories:

1. Suspension type
2. Strain type

5.7.1 Suspension type string

In suspension-type insulators, discs are placed horizontally, and the string is placed vertically as shown in Figure 5.12. These types of combinations are used in straight transmission line.

Figure 5.12 Suspension-type string of disc insulators

Based on design, three categories of discs are used in suspension-type application as follows:

- Interlinking type (also known as Hewlett type)
- Cemented cap type
- Cap and pin type
- Core and link type

Separation of vertical distance between above two consecutive discs are kept around 146 mm, 120 mm and 162 mm for interlinking type, cemented cap type and core and link type, respectively.

5.7.2 Strain type string

In strain-type insulators, discs are placed vertically, and the string is placed horizontally as shown in Figure 5.13. Two strings of disc insulators have been connected with cross arm of angle tension tower of 132 kV transmission line at one end and at other end, they are connected with line and jumper wires (Figures 5.13 and 5.14). These types of combinations are used at terminals and at the bending of transmission line.

5.8 Voltage distribution in string

OHL insulators act as capacitor. Series-connected capacitors show uniform voltage distribution. Though a string of insulators looks like series combination of capacitors, it does not show uniform voltage distribution. In the metal contact (node point) between any two disc and metal support of tower, there is air as dielectric medium and tower and the nodal points form capacitors through which capacitive current flows. Because of these paths, voltage distribution across the disc is not uniform. Capacitor equivalent network of string of n number of discs has been shown in Figure 5.15.

Figure 5.13 Strain-type string of disc insulators connected with cross arm of 132 kV transmission line

Figure 5.14 Use of both disc and pin on the same tower at 33 kV transmission line

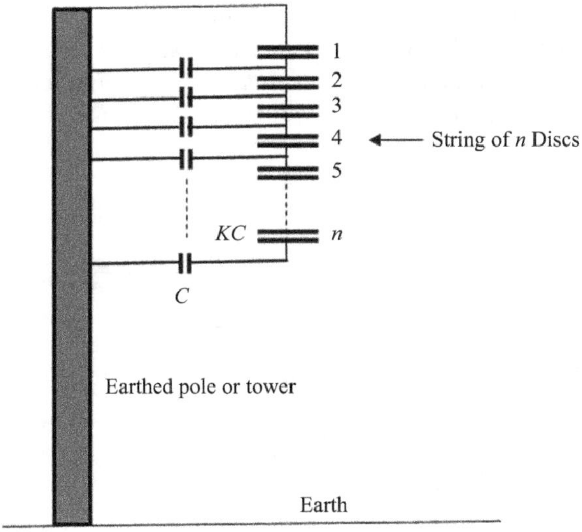

Figure 5.15 Capacitor equivalent network of string of 'n'-number of discs

Let, KC be equivalent capacitance of disc, C be capacitance between node and tower

$$I_2 = I_1' + I_1 \tag{5.1}$$

$$\omega CV_2 = \omega CV_1 + \omega KCV_1 \tag{5.2}$$

$$V_2 = (1 + K)V_1 \tag{5.3}$$

Therefore,

$$V_2 \neq V_1 \tag{5.4}$$

Thus, voltage distribution in string is not uniform. This shows that V_2 is greater than V_1. Voltage across the disc nearer to the conductor is maximum and nearer to the tower is minimum.

For a string consisting of n number of disc,

$$V_n > V_{n-1} > V_{n-2} > \ldots\ldots > V_3 > V_2 > V_1 \tag{5.5}$$

5.9 Effect of unequal voltage distribution

Voltage distribution across different discs in a string is not the same. Voltage across the disc nearer to the tower is less and nearer to the conductor is more. This makes 'continuous voltage stress' more at the discs which are located near conductor and increases probability of their failure. Addition to that overvoltage caused by lightning or other causes makes them more vulnerable.

Voltage (V_n) across different insulators of a string varies in steps with the increase of respective insulator number (n), as shown in Figure 5.16. It is smaller near the tower and increases towards conductor end. Change of voltage (V_n) along

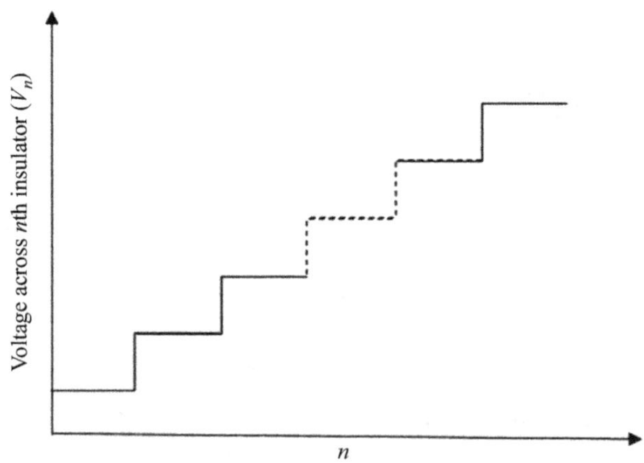

Figure 5.16 Voltage (V_n) across different insulators of a string versus respective insulator number (n)

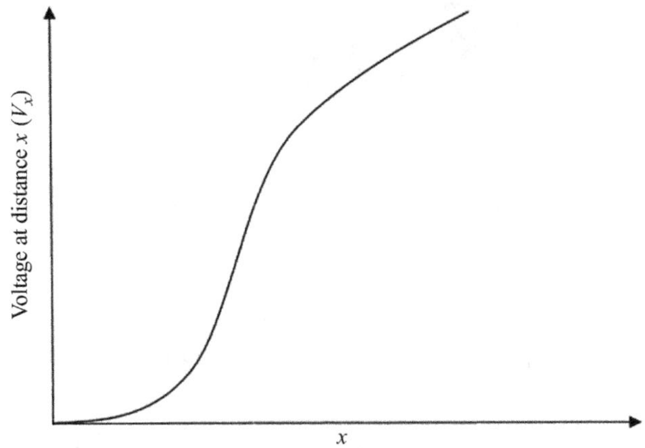

Figure 5.17 Voltage (V_x) at distance x versus x of composite silicone insulator for EHV line

the composite silicone insulator with the increase of distance from the tower cross end (x) has been shown in Figure 5.17. Instead of step change, it shows gradual change of voltage from tower end towards conductor end.

5.10 String efficiency

String efficiency (S_E) is defined as follows:

$$\text{String efficiency} = \frac{\text{voltage across the string}}{\text{voltage across the disc nearer to the conductor multiplied by the number of disc}}$$

or,

$$\text{String efficiency} = \frac{\text{voltage across the string}}{\text{maximum disc voltage in string multiplied by the number of disc}}$$

$$\text{String efficiency} = \frac{V_n + V_{n-1} + V_{n-2} + \ldots\ldots + V_3 + V_2 + V_1}{nV_n} \quad (5.6)$$

As

$$\begin{aligned} &V_n > V_{n-1} > V_{n-2} > \ldots\ldots > V_3 > V_2 > V_1 \\ &nV_n > (V_n + V_{n-1} + V_{n-2} + \ldots\ldots + V_3 + V_2 + V_1) \end{aligned} \quad (5.7)$$

$$\frac{V_n + V_{n-1} + V_{n-2} + \ldots\ldots + V_3 + V_2 + V_1}{nV_n} < 1 \quad (5.8)$$

String efficiency (S_E) will be less than 1.

Figure 5.18 Parallel string in 132 kV line

5.11 Improvement of voltage distribution and string efficiency

Improvement of voltage distribution across the discs of a string improves string efficiency. This reduces the unwanted effect of unequal voltage distribution. This is achieved in the following ways:

1. Connection of parallel string
2. Connection of multiple parallel composite silicone insulator
3. Connection of grading for guard ring
4. Connection of arcing horn

5.11.1 Connection of parallel string

Parallel strings are connected widely to improve string efficiency. Multiple numbers of strings are connected together at the end of cross arm and at the end of jumper and line conductor. The combination can serve the purpose, even after partial failure of few discs. Parallel string increases the lifetime of the line insulator system. Tension strings connected in parallel with a 132 kV line has been shown in Figure 5.18.

5.11.2 Connection of multiple parallel composite silicone insulators

Use of multiple parallel composite silicone insulators has become the first choice for extra high voltage (EHV) and ultra-high-voltage (UHV) overhead lines (OLHs). Parallel combination of four number of composite silicone insulator paths used in 765 kV transmission line has been shown in Figures 5.19 and 5.20. It assumes high lifetime which is yet to be tested. It offers comparatively high mechanical strength at light weight in addition to the other advantages of using composite material as OHL insulator.

5.11.3 Connection of grading for guard ring

Connection of grading for guard ring is an old but low-cost easy practice followed from long back. Rings are connected near the conductor end surrounding the string

Figure 5.19 Strain-type parallel strings in 765 kV line connected with cross arm (without jumper under construction)

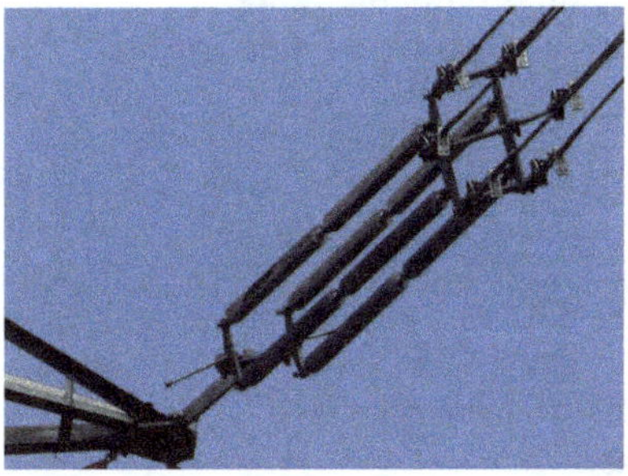

Figure 5.20 Strain-type parallel strings in 765 kV line (with bundled conductors under construction)

and conductor. This makes voltage gradient uniform and contributes additional capacitance with the nodes of string improving overall string efficiency.

5.11.4 Connection of arc horn

Connection of arc horn at the two ends of string serves two purposes. It protects from surge or overvoltages as well as improves string efficiency. Use of arc horn has also been a common practice in both distribution and transmission lines. Arc horn connected with string has been shown in Figure 5.21.

5.12 Selection practice for insulator

Selection of line insulator is done on the basis of the following criteria:

1. Operating voltage level
2. Pole type

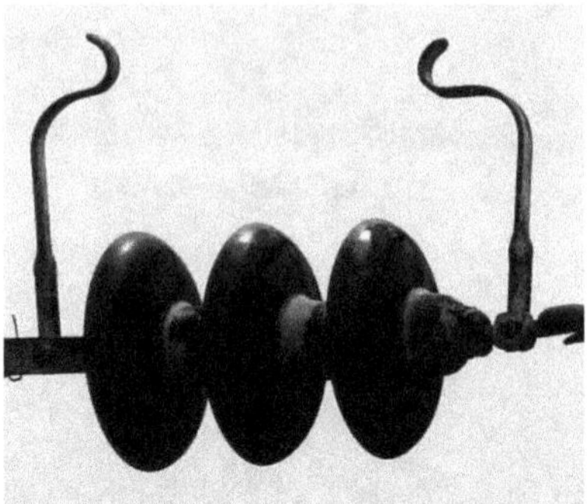

Figure 5.21 Arc horn connected with 33 kV line

3. Tower type
4. Cross arm arrangement
5. Location in the transmission path
6. Support height, sag and clearance requirement
7. Climatic condition
8. Wind speed statistics
9. Tower height
10. Pollution level
11. Stress level by conductor
12. Corona
13. Probable voltage distribution on string
14. Cost and availability

5.13 Associated design factors of insulators

Installation of OHL insulators are involved with design and installation of other subparts of the line. They are as follows:

1. Type of connection of insulator with support
2. Type of interconnection
3. Clamp and hold arrangement
4. Socket and ball arrangement
5. Grading or guard ring
6. Arc horn
7. Jumper wire
8. Conductor type and specification
9. Conductor binding/holding arrangement
10. Insulator lifting arrangement

5.14 Clamps

Clamps are used to fit insulators with cross arms or poles directly. Major desired properties of clamps may be written as follows:

- Easy to couple
- Damage proof oxidation proof
- High durability
- High mechanical strength
- High heat withstand capacity
- High humidity withstand capacity

In common practice, galvanized clamps are used. Different types of clamps are seen; some are as follows:

- Clamp with nuts and bolts.
- Number of nut and bolt combination may be one or more than one.
- Clamp with hook: clamp enclosed cross arm or pole body, and hook is used to connect the equipment with clamp.
- Rectangular-shaped clamp: these are useful to enclose rectangular pole body or any other member section.
- D-shaped clamp: useful for shackle-type insulator. It holds the insulator from its two sides.
- Compression-type dead-end clamp: this is used with composite conductor like ACSR (Aluminium conductor steel-reinforced) at dead-end support of distribution or transmission OHLs. It connects line conductor in one direction and jumper in other direction. It consists of steel sleeve compression unit to hold core of the conductor with tight gripping.
- Other shapes.

5.15 Nuts and bolts

These are small units in support facilities but important. Following should be kept in mind that they

- must have high mechanical strength,
- must have high heat and humidity withstanding capacity,
- should be galvanized,
- must be used with proper dimension as specified in particular application,
- may be used with washer and
- may be used with locks.

5.16 Failure of insulator

Causes of failure of insulators may broadly be divided into the following categories:

- Electrical causes
- Mechanical causes
- Thermal causes
- Ageing effect
- Others causes

5.16.1 Electrical causes

Major electrical failures are as follows:

1. Flashover
2. Short circuit

Overvoltage is major electrical cause for insulator failure. OHLs experience lightning and other surge voltage. Overvoltages may be divided into two categories: unidirectional and alternating. Lightning is unidirectional, whereas switching surges propagated through the line are alternating (power frequency and high frequency). Also corona effect causes high voltage gradient in the line insulators.

Pin broken on the upper side at 33 kV line has been shown in part (a), while pin broken at lower end at 33 kV line has been shown in part (b) of Figure 5.22.

Figure 5.22 (a) Pin broken on the upper side at 33 kV line, (b) pin broken at lower end at 33 kV line, (c) damaged string of discs at 33 kV distribution line, (d) damaged disk in 11 kV distribution line

5.16.2 Mechanical causes

Excessive pressure due to unbalanced conductor tension may damage the overall structure or in partial of line insulator.

High wind flow and ice contribute to unbalanced pressure on the insulator. This may damage near the conductor end or near the clamp and hold area.

Some examples of insulator failures have been shown in Figure 5.23. Tilted pin of 33 kV line due to wind pressure (cyclone: Amphan, West Bengal, India 2020) has been shown in Figure 5.23(a), torn insulator joint at 440 V line has been shown in Figure 5.23(b).

Major mechanical failure may be listed as follows:

1. Mechanical over and unbalanced stress
2. Cracks
3. Improper coupling or jointing
4. Porosity
5. Dust deposition

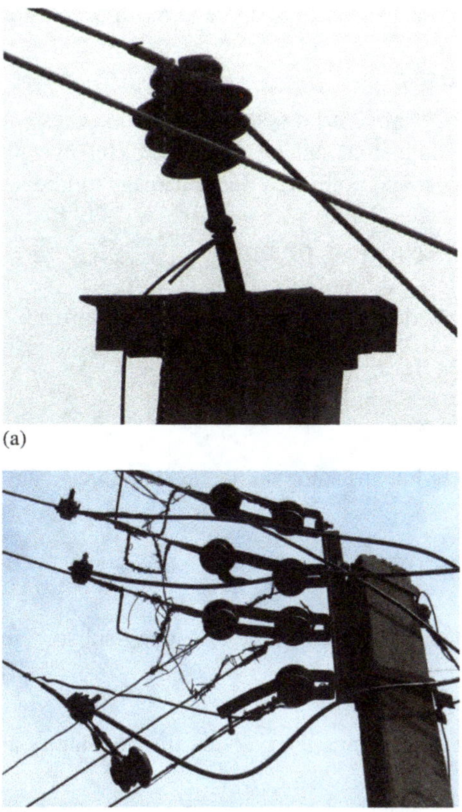

(a)

(b)

Figure 5.23 (a) Tilted pin of 33 kV line due to wind pressure (Amphan, West Bengal, India, 2020), (b) torn insulator joint at 440 V line

5.16.3 Thermal causes

Capability of withstanding different climatic condition is not the same for all types of OHL insulators. Thermal distribution over long period depends on climatic situation. Therefore, changes in the extreme thermal behaviour weather have effects on insulating material. Changes occur slowly; however, sometimes, rapid change in temperature along with humidity is also experienced. Abnormal changes may result in unpredictable failure of line insulators.

5.16.4 Ageing effect

Though porcelain and grass OHL insulators have shown long lifespan, even then there is the possibility of quality fall of materials. This may change leakage and charging current characteristics of the insulator. Also accumulated dust spot and degradation of polishes of the surface over ageing is observed. Dust accumulation may be a problem and regular cleaning may be a challenge for composite silicone insulator. Regular inspection should be carried out. Photography, helicopter and application of unmanned aerial vehicle may be subject area for further development in this area. Composite silicone material is expected to have less affected by climatic condition; however, it is yet to be tested in real-life practice; therefore, time has not come to make conclusion.

5.16.5 Other causes

Failure of other parts of the overhead system may be the cause of insulator failure. For example, conductor failure, tree fall on conductor, etc. may damage line insulator. Structural failure of tower and poles may cause damage to line insulators (Figure 5.24).

5.17 Standards, test and practice

OHL insulator property depends on the following features:

1. Electrical features
2. Mechanical features
3. Thermal features

Tests carried out for line insulator are divided into three main categories as follows:

1. Routine test
2. Performance test
3. Flashover test

All these have been discussed in the following subsections.

5.17.1 Electrical features

Major electrical properties characterizing the line insulators are as follows:

1. Capacitance
2. Insulation resistance
3. Dielectric dissipation factor (DDF) and power factor (PF)

Figure 5.24 Ruptured string

5.17.1.1 Capacitance

Capacitance of a line insulator generates from the interaction between electrical property permittivity and its geometry. Mathematically, capacitance of a line insulator can be expressed as follows:

$$C = K_G \varepsilon \tag{5.9}$$

where, C is the capacitance, ε is the permittivity of insulator material, and K_G is the constant dependent upon geometry of the insulator.

Now,

$$\varepsilon = \varepsilon_r \varepsilon_0 \tag{5.10}$$

where, ε_0 is the permittivity of free space and ε_r is the relative permittivity.

Therefore,

$$C = K_G \varepsilon_r \varepsilon_0 \tag{5.11}$$

5.17.1.2 Insulation resistance

From definition, resistance may be expressed as follows:

$$R = \rho \frac{l}{A} \tag{5.12}$$

where ρ is the resistivity of the insulator material, l is the length, and A is the cross section of the insulator.

Resistance of an insulator is defined and measured as per the geometry of the insulator by volume resistance and surface resistivity.

Volume resistance

Volume resistance of line insulator refers to the resistance measures across insulator material of unit cross section (1 m^2) placed between two electrodes separated by 1 m length. It is expressed by Ohm-m (Ω m).

Surface resistance

Surface resistance of line insulators to the resistance offered by surface insulants of unit area is measured between two exact opposite sides of insulator. It is also expressed by Ω m.

5.17.1.3 Dielectric dissipation factor and power factor

Dielectric dissipation factor (DDF) refers to the index of rate of loss of energy of an insulator. Let us consider that due to V voltage across insulator, current is I. Current will have two components: loss equivalent resistive current component (I_r) and charging and discharging current component (I_c).

Let, ∅ be the angle between I_r and I or I and V; δ is the angle between I_c and I. DDF is written as follows:

$$DDF = \tan\delta = \frac{I_r}{I_c} \tag{5.13}$$

Power factor (PF) can be mathematically expressed as follows:

$$PF = \cos \varnothing = \frac{I_r}{I} \tag{5.14}$$

5.17.1.4 Time constant (discharging–charging property)

Time constant of an insulator is an important property which describes the rate of charging and discharging. Time constant is assessed by applying pulse of direct voltage across the insulator. Rise of voltage across the insulator is observed. Time of occurrence of peak voltage is noted. It is then discharged and again charged, and time of recovery of peak voltage is noted. The difference between these two noted times gives the measure of time constant.

5.17.1.5 Partial discharge

Partial discharge refers to very localized electrical discharge through insulating material or surface. Often it is found that partial discharge is associated with either or all of light, sound, heat and chemical reaction along with the voltage of amplitude of few micro-volts to volt. Strong electrical stress for very short period of time (less than 1 s) which is greater than dielectric strength causes partial discharge (IEC-60270). Size of the composite filler particle plays an important role in a composite insulator. Resistance to partial discharge increases with the decrease of filler size. Partial discharge depends on dielectric strength of the material which is dependent on DDF, volume resistivity and surface resistivity of the insulator.

5.17.1.6 Erosion resistance

Erosion resistance refers to the loss of weight of insulating material of an insulator. Screening is standard procedure for the investigation of erosion.

5.17.1.7 Tracking resistance

Tracking resistance refers to the path formation by carbon, dust, etc. on the surface material of an insulator. Screening is also standard procedure for the investigation of tracking resistance.

5.17.1.8 Resistance to penetration of water

Penetration of water through the surface for porcelain or glass material or through silicone rubber in composite silicone insulator is an important aspect which decides their usage as OHL insulator. The higher the resistance to this path, the better the performance is. The absence of water penetration is the perfect state of condition. Water penetration causes decrease of dielectric strength and may result in severe breakdown.

5.17.2 Mechanical features

Major mechanical properties characterizing the line insulators are as follows:

5.17.2.1 Corrosion resistance

Corrosion resistance is the virtue of the insulating material which opposes corrosion process. More is the value of corrosion resistance, slower is the corrosion rate.

5.17.2.2 Tensile stress

Tensile stress refers to load per unit area of the insulator.

5.17.2.3 Elongation

Increase in length due to tensile stress is known as elongation.

5.17.2.4 Ductility

Maximum elongation before reaching rapture is known as ductility.

5.17.2.5 Strain

Increase in length due to tensile stress with respect to the original length of test material is known as strain.

5.17.2.6 Percentage of elongation

Percentage increase in length due to tensile stress with respect to the original length of test material is known as percentage elongation.

5.17.2.7 Elasticity modulus

Ratio of stress to strain within linear relationship is known as elasticity modulus.

5.17.2.8 Hardness

Hardness refers to the property of the insulating material which opposes scratch, indentation, wear, abrasion, etc.

5.17.2.9 Impact strength

Energy required per unit area of cross section of the test material to fracture is known as impact strength.

5.17.2.10 Yield point

Yield point refers to the point where strain changes without change of stress in stress–strain characteristics.

5.17.2.11 Wear resistance

Wear resistance is the property of the insulating material which opposes changes in physical shape or dimension during sliding of surface of other body.

5.17.3 Thermal features

Major thermal properties characterizing the line insulators are as follows:

5.17.3.1 Thermal conductivity

Let us consider the following variables during the flow of heat through test materials:

Q is the amount of heat flow (kcal)
L is the length of heat flow in metre
A is the cross-sectional area in square metre through which heat flows
T is the temperature difference in centigrade scale
t is the duration of heat flow
Then,

$$\begin{aligned} Q &\propto A \\ Q &\propto T \\ Q &\propto t \\ Q &\propto \frac{1}{L} \end{aligned}$$

Therefore,

$$Q \propto \frac{ATt}{L} \tag{5.15}$$

$$Q = K\frac{ATt}{L} \tag{5.16}$$

where, K is the proportionality constant.

$$K = Q\frac{L}{ATt} \tag{5.17}$$

If,

$$\begin{aligned} L &= 1 \\ A &= 1 \\ T &= 1 \\ t &= 1 \end{aligned}$$

Then,

$$K = Q \tag{5.18}$$

K is known as 'thermal conductivity'. It may be defined as the heat flow through unit distance of the material of unit temperature difference through unit cross-sectional area for unit duration.

5.17.3.2 Thermal resistance

Thermal resistance is the property of the material by which it opposes flow of heat through it.

We know that,

$$Q = K\frac{ATt}{L}$$

where, Q is the amount of heat flow (kcal), L is the length of heat flow in metre, A is the cross-sectional area in square metre through which heat flows, T is the temperature difference in centigrade scale, t is the duration of heat flow, and K is known as thermal conductivity.

Therefore, rate of heat flow may be written as follows:

$$H = \frac{Q}{t} = K\frac{AT}{L} \tag{5.19}$$

or,

$$\frac{T}{H} = \frac{L}{AK} \tag{5.20}$$

This is analogous with the flow of charge and can be compared with the relation between the current (I) flow due to voltage difference (V) as follows:

$$\frac{V}{I} = R \tag{5.21}$$

where, R is the electrical resistance of the path.

Comparing the previous two equations, thermal resistance may be written as follows:

$$R_{Thermal} = \frac{T}{H} = \frac{L}{AK} \tag{5.22}$$

5.17.3.3 Specific heat

An amount of heat required to increase temperature 1 °C of unit mass of a material is known as specific heat (s).

5.17.3.4 Heat capacity

Heat capacity of a material refers to the specific heat of the material multiplied by the mass of the material under consideration. Mathematically, heat capacity (S) may be written as follows:

$$S = sm \tag{5.23}$$

where, s is the specific heat and m is the mass of the material.

5.17.3.5 Thermal expansion

Thermal expansion refers to the increase in dimension (length, area and volume) due to the increase of temperature.

5.17.4 Routine test

OHL insulators once installed need regular monitoring and routine health check-up. Small failure in insulator cannot be neglected, as it may lead to severe failure. Repair of some parts of insulator can make the system run for a considerable time. But, in most of the cases, repair is not a solution; rather, replacement of part or whole of the insulation installed in support line is followed. Different types of routine tests are as follows:

1. Mechanical test
2. Corrosion test
3. High-voltage test

Mechanical test: Material used in line insulator has its own stress–strain characteristics. After getting particular shape as insulator and way of connection, OHL characteristics should well match with conductor tension, wind and ice effect. In mechanical test, around 25% over-tension is applied and changes are observed. Mechanical test is an important routine test to be carried out for safety and longer life of the line insulator.

Corrosion test: Overhead insulator arrangement is undergone corrosion test. This is done by immersing the system in chemical solution of copper sulfate. Test duration is set for around 1 min at 15 °C. Then it is removed and test is repeated several times. Depositions on the surface are monitored.

High-voltage test: This test is performed by applying high voltage across the insulators. Both high-voltage AC and DC are used for testing [2]. Mainly pin-type insulators are tested in this way. Pins are immersed partially in water in reversed direction, and high voltage is applied to see the effect.

5.17.5 Performance test

Different types of routine tests are as follows (Table 5.2):

1. Puncture test
2. Porosity test
3. Mechanical test
4. Thermal test
5. Electromechanical test

Puncture test: Line insulator is immersed in insulating oil and voltage is applied across it. Voltage is then gradually increased. It is increased until puncture occurs and the voltage is noted. Puncture voltage of a string is kept higher than flashover voltage at wet condition; more than 30% higher value than flashover is set as puncture voltage.

Table 5.2 Different types of test practiced for overhead line insulators

Type test		Routine test	Optional routine test	Sample test	Optional sample test
Standard	**Optional**				
Voltage withstand at power frequency (wet and dry)	Impulse voltage withstand test	Visual inspection	Ultrasonic inspection	Locking verification	Impulse voltage puncture test
		Electrical test		Displacement verification	
Voltage withstand at impulse frequency (dry)	Zinc sleeve test	Mechanical test		Thermal cycle test	Zinc sleeve test
				Porosity test	
Dry fast front withstand test	Residual strength test			Electromechanical test	Radio interference voltage (RIV) test
Thermal performance				Puncture test	
Mechanical performance				Thermal shock test	
Electromechanical failing				Galvanizing test	
Dimension verification					

Porosity test: This test is carried out with the help of fuchsine dye. First solution of fuchsine dye is made in alcohol. An OHL insulator sample is taken and broken in pieces. The broken pieces are immersed in the solution at high pressure for several hours and then removed. Penetration of dye into the sample under test is observed.

Mechanical test: One pin-type insulator is placed above another rigidly fixed pin. Wire is encircled around the groove. Load is applied perpendicularly and on the wire. An amount of load is increased gradually. It is seen whether there is any breakage occurs below minimum specified failing load.

Temperature cycle thermal test: An insulator is immersed on hot and cold water periodically for equal durations. The same test is repeated for several times and it is checked whether there are cracks occurred on the glazed surface. This test is also known as temperature cycle test.

Electromechanical test: In this test, both electrical and mechanical stresses are applied on an insulator. Normally, suspension-type insulators undergo this type of test. The test starts with the application of around 75% dry flashover voltage and 50% of failure stress. Then, gradually both of them are increased and it is observed whether there is any breakage or puncture occurred in the insulator.

5.17.6 Power test

Different types of routine tests are as follows:

1. Power frequency test: dry test and wet test
2. Impulse frequency test

Power frequency test: Power frequency test is performed in dry condition as well as in wet condition.

In dry test, an insulator(s) is placed in dry atmospheric condition. One of ten numbers of insulators (separated by a distance equal to four times of individual length) is arranged in the way as they will be used. Voltage is applied across the test sample(s) and then gradually increases until and unless flashover takes place. Voltage is noted as dry flashover voltage.

In wet test, the same experiment is carried out by spraying water on the surface. Like in dry test, an insulator(s) is placed in dry atmospheric condition. One of ten numbers of insulators (separated by a distance equal to four times of individual length) is arranged in the way as they will be used. Voltage is applied across the test sample(s) and then gradually increases until and unless flashover takes place. Voltage is noted as wet flashover voltage.

Both the experiments are repeated several times and average result is considered.

Impulse frequency test: Impulse frequency high-voltage generator is used to carry this test. Impulse voltage (in the order of several hundred kHz) is applied across the insulator under test. Flashover voltage is then noted.

Impulse ratio: When both power frequency and impulse frequency flashover tests are conducted, impulse ratio is also calculated. It refers to the ratio of flashover voltage obtained by impulse test to the flashover voltage obtained by wet power frequency test. Thus, it may be written as follows:

$$\text{Impulse ratio} = \frac{\text{flashover voltage obtained by impulse frequency flashover test}}{\text{flashover voltage obtained by power frequency flashover test}}$$

As there are two types of power frequency tests, there are two impulse ratios considering the ratio of impulse frequency flashover voltage with dry test and wet test flashover voltage.

5.17.7 Practice tests

While installing line insulators in particular OHL, the following practices are follows:

1. Type test
2. Sample test
3. Routine test

5.17.8 Effect of environment on leakage current

Leakage path forms over the surface of an insulator. It varies with the change of type of insulator. Therefore, current in the leakage path is not the same and varies with the change of type of insulators. Moreover, different environmental conditions have influence over leakage current. In clear weather condition, Class E insulator shows minimum leakage current, whereas Class A insulator shows greater leakage current. Rain increases leakage current in all the classes of insulators, approximately 7–16 times than clear weather condition. During fog, it increases 10–25 times and in snow, it increases 14–75 times. Therefore, leakage current withstanding sample tests for different weather conditions must be carried out.

5.17.9 Ageing effect

Like other sections of the power system, OHL insulators also undergo through ageing effect. Porcelain and glass materials are used as line insulator for more than hundred years, and their performances have been experienced. On the other hand, composite silicones are also performing well in high-voltage lines under different climatic conditions. Normal operating condition and suitable climatic conditions, question on lifespan of ageing will not appear. But, when an OHL undergoes different (normal and abnormal disturbances) operating conditions and wide variations in climatic neighbours, question on ageing effect will appear. But, it is a very difficult and changing task to a monitor, which can be concluded on the ageing effect under running or operating conditions of the transmission line. Some ageing effects of OHL insulators include degradation of surface property, corrosion, etc. Lots of researches are going on to study the ageing effects of line insulators. Some following techniques are being adopted for monitoring ageing effect from outside at running condition:

- Photography with light camera
- IR camera with night vision
- Helicopter application
- Unmanned aerial vehicle application

In test process, one should have clear idea about the standard features of the object and the test procedure to be followed. Analysis to reach the conclusion available is mostly numerical approaches.

5.18 Summary

Different aspects of insulators used in OHLs have been described in this chapter. Materials used in insulators have been discussed. Different design features and factors deciding the selection of OHL insulators have been discussed. String

Table 5.3 Useful standards and guidelines

SN	Standard [3–17]	Area of interest
1	BS 137	Specifications of insulators of ceramic material or glass for overhead lines with a nominal voltage greater than 1,000 V
2	CIGRE SC22 WG22-03	Service performance of composite insulators on HVDC lines Protective device for insulator sets (HV and EHV transmission overhead lines)
3	CISPR 18-3	Radio interference characteristics of overhead power line and high-voltage equipment. Part 3: code of practice for minimizing the penetration of radio interference
4	EN 50 186	Live-line washing systems for power installation with nominal voltages above 1 kV
5	EN 60 433	Insulators for overhead lines with a nominal voltage above 1 kV, ceramic insulators for AC systems – characteristics of insulator units of the long-rod type
6	EN 61 325	Insulators for overhead lines with a nominal voltage above 1,000 V – ceramic or glass insulator units for DC systems – definitions, test methods and acceptance criteria
7	IEC 60 050-471	International electrotechnical vocabulary
8	IEC 60 120	Dimensions of ball and socket couplings of string insulator units
9	IEC 60 305	Insulators for overhead lines with a nominal voltage above 1 kV ceramic or glass insulator units for AC systems – characteristics of insulator units of the cap and pin types
10	IEC 60 437	Insulators for overhead lines with nominal voltage above 1,000 V. AC power arc test on insulator sets
11	IEC 60 507	Artificial pollution test on high-voltage insulators to be used on AC system
12	IEC 60 671	Insulation coordination
13	IEC 60 672	Ceramic and glass insulating materials
14	IEC 60 720	Characteristics of line post insulators

efficiency has been determined. Causes of failure of OHL insulators have been discussed. Tests and failure for insulators have been presented. At the end, useful standards have been presented for further study.

5.19 Useful standards and guidelines for further study

Some useful standard guidelines resources for further study of OHL insulators have been presented in Table 5.3.

References

[1] H. Khan, M. Amin, and A. Ahmad, 'Characteristics of silicone composites for high voltage insulations'. in Reviews on Advanced Materials Science, vol. 56, pp. 91–123, 2018.

[2] B. M. Bailey, 'Test line experience with HVDC overhead transmission'. in IEEE Transactions on Power Apparatus and Systems, vol. PAS-89, no. 7, pp. 1625–1634, 1970.

[3] BS 137: Specifications of Insulators of Ceramic Material or Glass for Overhead Lines With a Nominal Voltage Greater Than 1000 V. Part 2: Requirements. BSI Committee - PEL/36, London; 1973.

[4] 'CIGRE SC22 WG22-03: service performance of composite insulators on HVDC lines'. in Electra, vol. 191, pp. 53–57, 2000.

[5] 'CIGRE SC22 WG22-03: protective device for insulator sets (HV and EHV transmission overhead lines)'. in Electra, vol. 136, pp. 22–29, 1991.

[6] Comité International Spécial des Perturbations Radioélectriques (CISPR): Radio Interference Characteristics of Overhead Power Line and High Voltage Equipment. Part 3: Code of Practice for Minimizing the Penetration of Radio Interference. IEC, Geneva; 1986.

[7] EN 50 186: Live-Line Washing Systems for Power Installation With Nominal Voltages Above 1 kV. CENELEC, Brussels; 1999.

[8] EN 60 433: Insulators for Overhead Lines With a Nominal Voltage Above 1 kV, Ceramic Insulators for AC Systems – Characteristics of Insulator Units of the Long Rod Type. CENELEC, Brussels; 1998.

[9] EN 61 325: Insulators for Overhead Lines With a Nominal Voltage Above 1000 V – Ceramic or Glass Insulator Units for DC Systems – Definitions, Test Methods and Acceptance Criteria (IEC 61325). CECELEC, Brussels; 1995.

[10] IEC 60050-471: International Electro-Technical Vocabulary, Insulator. IEC, Geneva; 1984.

[11] IEC 60 120: Dimensions of Ball and Socket Couplings of String Insulator Units. IEC, Geneva; 1987.

[12] IEC 60305: Insulators for Overhead Lines With a Nominal Voltage Above 1 kV Ceramic or Glass Insulator Units for AC Systems-Characteristics of Insulator Units of the Cap and Pin Type. CENELEC, Brussels; 1996.

[13] IEC 60437: Insulators for Overhead Lines With Nominal Voltage Above 1000 V. AC Power Arc Test on Insulator Sets. IEC, Geneva; 1997.
[14] IEC 60 507: Artificial Pollution Test on High Voltage Insulators to be Used on AC System. IEC, Geneva; 1991.
[15] IEC 60 671: Insulation Coordination. IEC, Geneva; 1996.
[16] IEC 60 672: Ceramic and Glass Insulating Materials. IEC, Geneva; 1999.
[17] IEC 60 720: Characteristics of Line Post Insulators. IEC, Geneva; 1981.

Chapter 6

Conductor

This chapter deals with conductor used in overhead electric lines. It describes electrical, thermomechanical and other properties of line conductors. The conductors are classified followed by their description. Stranded and bundled conductors are described. Materials used in conductor have been presented. Conductors of the same material and composite conductor have been described. Different aspects of hollow conductor, conductor with optical fibre cable, phase and earth conductors have been presented. Different types of jumper have been mentioned. Covered conductors or overhead cables have been included. Conductor fittings and installation care have been described. Common stringing methods followed by different electrical properties have been discussed. Ground clearance, stringing chart, conductor vibration, damper, spacer, etc. have also been presented.

6.1 Introduction

The main purpose of any overhead lines is to carry electric power from one end to another through conductor. Therefore, electrical feature is important. In addition to this, conductor strength and tension have impact on the support structure of the overhead lines. Therefore, physio-thermal and mechanical properties must also be considered in the study of overhead lines. With the advancement of technology, different types of conductors have been introduced for overhead lines. Conductors made of single material as well as composite materials are found in practice. Different features of overhead line conductors have been discussed in this chapter.

6.2 Conductor property

Conductors in overhead lines are the materials which carry electric power from one end to another. As the name suggests, this should be made of conducting materials as major part and should have high conductivity. Conductors are described by the following two categories of properties as follows:

- Electrical properties
- Thermomechanical properties

Conductors used in the overhead lines are mostly bare conductors. Use of covered conductors has started in low-voltage distribution lines. Conductors used in overhead lines are stranded and used in single or in bundled condition.

6.2.1 Electrical properties of line conductors

Electrical properties [1,3] of line conductors that should be considered for their design, selection and installation are as follows:

- DC resistance (Ω/km)
- Current carrying capacity or ampacity (A) or steady-state thermal rating or thermal rating: maximum amount of current that conductor can carry maintaining the security and safety of the conductor
- Temperature coefficient of resistance (1/K): rate of change of resistance with respect to change of temperature
- Specific resistivity ((Ω mm^2)/m)
- AC resistance

6.2.2 Thermomechanical properties of line conductors

Mechanical properties of line conductors that should be considered for their design, selection and installation are as follows:

- Density (kg/mm^3)
- Specific conductivity (m/Ω mm^2)
- Specific heat (Ws/GK)
- Coefficient of thermal expansion (10^{-6}/K)
- Modulus of elasticity (kN/mm^2)
- Tensile stress at 1% elongation (N/mm^2)
- Tensile stress before stranding (N/mm^2)
- Percentage elongation (%)
- Every day stress (N/mm^2)
- Cross-sectional area (mm^2)
- Number of strands (unit less)
- Strand diameter (mm)
- Conductor diameter (mm)
- Rated strength (kN)
- Breaking strength (kN)
- Mass per unit length (kg/km)
- Knee-point temperature (°C)
- Maximum allowable operating temperature (°C)
- Conductor-manufactured length (m)
- Steel area ratio
- Lay length (m)
- Lay ratio

6.2.3 Stranded conductors

Overhead line conductors made of wires of the same or different materials are made stranded. This improves mechanical strength as well as helps in reducing corona, skin effect, etc. Stranded conductor has been shown in Figure 6.1.

Figure 6.1 Stranded conductor

6.2.4 Bundled conductors

Conductors are the main components which carry electric load. Increase of electrical load carrying capacity by increasing diameter is not a feasible solution in bulk amount of power transmission and distribution. Instead of increasing diameter excessively, a bundled conductor is used. It increases current carrying capacity or ampacity without changing individual conductor dimension and hence keeping weight and conductor tension within the limit of withstanding capacity by the support and insulator arrangements.

Thus, in a bundled conductor, single conductor is replaced by multiple conductors as follows:

1. Two conductors
   i. Vertical
   ii. Horizontal
2. Three conductors
   i. Delta
   ii. Reversed delta
3. Four conductors (square)
4. Five conductors (pentagonal)
5. Six conductors (hexagonal)
6. Eight conductors (octagonal)

Conductors' position of bundled conductors of different shapes has been shown in Figure 6.2. Bundled conductor configuration of six sub-conductors used in 765 kV transmission line has been shown in Figure 6.3. Bundled conductor configuration of two sub-conductors used in 220 kV transmission line has been shown in Figure 6.4.

6.3 Materials

Materials chosen for conductor are mainly based on their properties and place of application. Depending upon choice and requirement, both single-type and

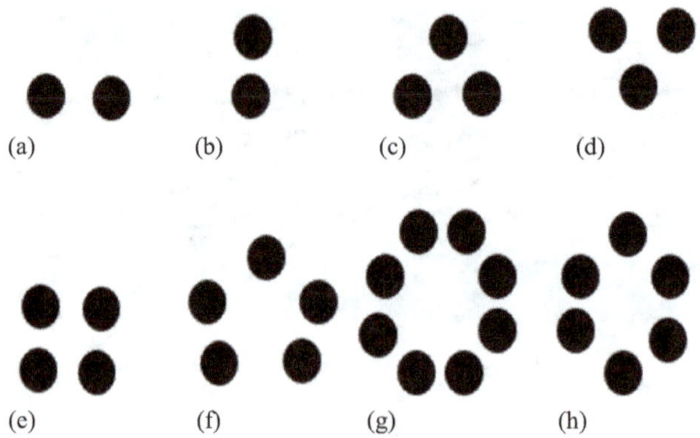

Figure 6.2 Conduct or positions of bundled conductors of different shapes (a) two conductor in horizontal position, (b) two conductors in vertical position, (c) three conductors in positioned in delta shape, (d) three conductors positioned in reversed delta shape, (e) four conductors positioned in square shape, (f) five conductors positioned in pentagonal shape, (g) eight conductors positioned in octagonal shape, (h) six conductors positioned in hexagonal shape

Figure 6.3 Bundled conductor configuration of six sub-conductors used in 765 kV transmission line

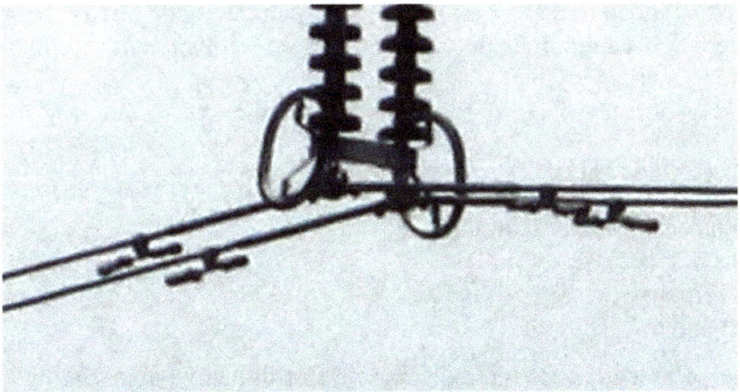

Figure 6.4 Bundled conductor configuration of two sub-conductors used in 220 kV transmission line

Table 6.1 Common feature of copper and aluminium

Property	**Copper**	**Aluminium**
Resistivity	1.68 μΩ cm	2.65 μΩ cm
Specific gravity	8.9 g/cm^3	2.71 g/cm^3
Melting point	1,085°C	660°C
Temperature coefficient of resistance at 20°C	0.00386/°C	0.00429/°C

composite-type materials are found suitable for application in overhead lines. Materials used in conductor in overhead lines are as follows:

- Aluminium
- Aluminium alloy
- Aluminium clad steel
- Copper
- Copper alloy
- Cadmium copper alloy
- Phosphor bronze
- Steel
- Coated or galvanized steel

Some common features of copper and aluminium have been compared as presented in Table 6.1.

Copper shows high conductivity and therefore its electrical properties are highly suitable for conductor. Aluminium has relatively less conductivity. But it has lesser weight which reduces sag of the line conductor. Melting point and specific gravity of aluminium are less than that of copper. Tensile strength of aluminium is also low; in high-voltage application, aluminium wires are used with steel core offering high mechanical strength with considerable electrical property. Aluminium is available plenty in nature (in the form of bauxite, etc.) and it is less-costlier than copper.

Cadmium added to copper alloy reduces conductivity by 5–10% but improves tensile strength. Cadmium copper alloy is used where high conductivity with high tensile strength is required.

Phosphor bronze offers the following properties:

- Good toughness property
- Low coefficient of friction
- High mechanical strength
- Grains of finer size
- Less corrosive
- Less fatigue

Phosphor bronze is used in transmission line in marine areas and in cryogenic applications.

6.4 Conductor types

Based on the materials, conductors are classified into the following two categories:

1. Conductors with the same material
2. Conductors with composite materials

6.4.1 Conductor of the same material

This type of conductors is made of single metal alone or by metal clad steel as shown in Figure 6.5. Thus, this type is again may be divided into the following two categories:

1. Monometallic
2. Metal clad steel

Metal clad steel conductor offers better mechanical strength than monometallic conductor. Some examples of conductor of the same material are as follows:

- All aluminium conductor (A1): made of aluminium
- All aluminium high-strength (HS) alloy conductor (A3): made of aluminium alloy

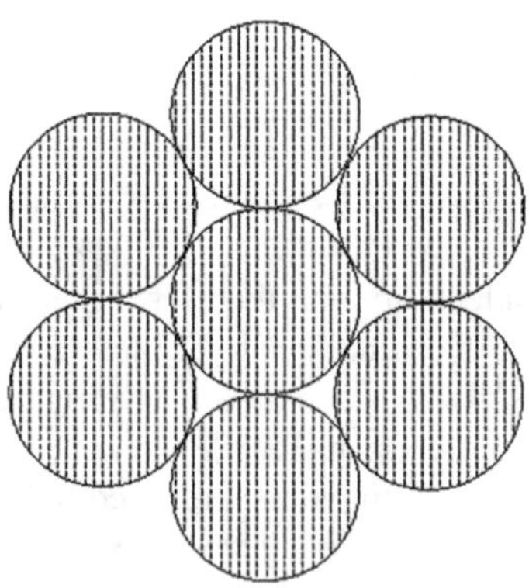

Figure 6.5 Conductor of same materials

- Aluminium conductor alloy reinforced (ACAR)
- Steel conductor: made of steel
- HS steel
- Extra HS steel
- Copper conductor: made of copper
- Aluminium clad steel: made of steel covered by uniform aluminium

In overhead lines, mainly the following types of conductor of the same materials are found:

- All aluminium conductor
- All aluminium alloy conductor

6.4.2 Composite conductors

Composite conductors are made of two or more number of materials. The structure of composite conductor is divided into two parts as shown in Figure 6.6. Inner part of the composite conductor is known as core. Core is surrounded by multiple numbers of layers. In most of the cases, core is made of steel. It offers high mechanical strength. Outer layers contribute to electrical conduction. Both core and outer layers are filled by wires of different shapes. With the advancement of material technology, wires of different shapes have been introduced for the manufacturing of conductors. Based on design features, different composite conductors have come into application [1–3]. Some examples are as follows:

- ACAR: aluminium conductor alloy reinforce
- ACSR: aluminium conductor steel reinforce

Numbers of wires placed in core are not the same with the number of wires placed in outer layers. If numbers of wires kept in outer layers and core are X and Y, respectively, as shown in Figure 6.7, then, the composite structure is known as X/Y composite conductor. Some different composite wire compositions with respect to the number of wires used in core and outer layers have been presented in Table 6.2. The total number of wires in conductor is $(X + Y)$. Some available diameters for aluminium wire are 3.78 mm, 4.14 mm, 4.44 mm, 5.06 mm, etc. Some available diameters for steel wires are 2.25 mm, 3.45 mm, 5.78 mm, 4.14 mm, etc. Some composite conductor diameters are 26.4 mm, 27 mm, 28.1 mm, 29 mm, 34 mm, etc.

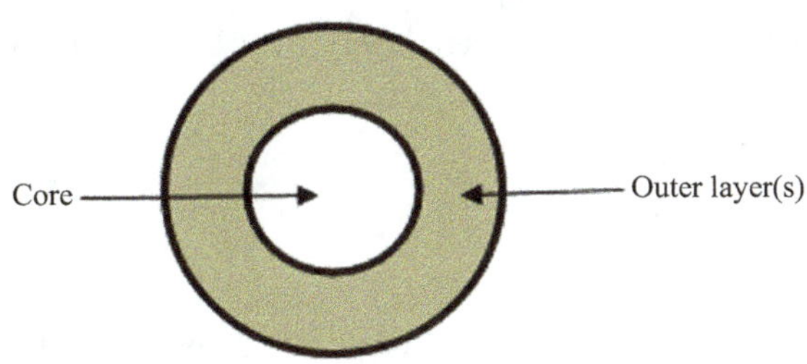

Figure 6.6 Two sections of composite conductor

Figure 6.7 Composite conductor filled by wires of two different materials

Table 6.2 Different composite wire compositions

Structure composition (X/Y)	Number of conducting aluminium wires at outer layers (X)	Number of steel wires at core (Y)
36/1	36	1
24/7	24	7
26/7	26	7
30/7	30	7
45/7	45	7
54/7	54	7
16/19	16	19
30/19	30	19
38/19	38	19
21/27	21	37

Shape of the wires placed in core and outer layers plays an important role in deciding properties of the conductor. Core wires are mostly seen to have circular shape. But shape of the wires used in outer layer may be circle or trapezoid or others as shown in Figure 6.8(a) and (b). In some cases, different outer layers have wires of different shapes as shown in Figure 6.8(c). Sample composite conductor used in 765 kV transmission line has been shown in Figure 6.9. Some composite conductors of different wire shapes have been presented in Table 6.3.

TW (Trapezoid-shaped wire) conductor: Conductor made of trapezoid aluminium strands are known as TW conductor.

The space between core and outer layers may be designed (as presented in Table 6.4) as follows:

- without any additional materials
- filled with other materials like zinc

6.4.2.1 Steel area ratio

Steel area ratio is an important parameter for indicating relative mechanical strength of composite conductor. It refers to the area of all stranded helical conductors to the area of the core. Sometimes it is simply called as area ratio.

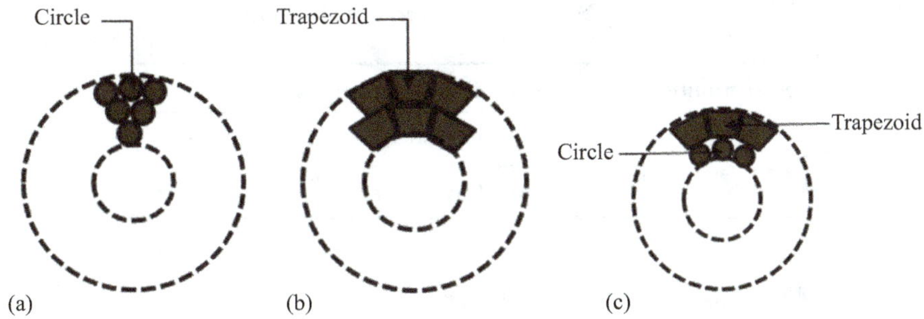

Figure 6.8 Composite conductor outer layer filled by wires of following shapes: (a) circle, (b) trapezoid, (c) circle and trapezoid

Figure 6.9 ACSR conductors used a 765 kV transmission line

Table 6.3 Some composite conductors of different wire shapes

Shape of wires in core	Shape of wires in outer layers
Circle	Circle
Circle	Trapezoid
Circle	Circle and trapezoid
Circle	Circle and other customize shape
Circle	Trapezoid and other customize shape
Circle or other combination	Other combinations

Table 6.4 Space between core and outer layers

Serial number	Space between core and outer layers
1	Filled by air
2	Filled by
3	Filled by others

6.4.2.2 Lay length and lay ratio

Lay length of a composite conductor refers to the one helical wire required to complete one turn of helix surrounding one layer. Lay ratio refers to the ratio of lay length to the outer diameter of the helix under consideration.

6.4.2.3 Knee-point temperature

Knee-point temperature refers to the temperature above which aluminium strand has no tension and starts compression.

6.4.2.4 Maximum allowable operating temperature

Maximum operating allowable temperature of overhead line conductor refers to the temperature above up to which it can carry electrical power load maintaining security and safety of the line.

6.4.2.5 Conductor-manufactured length

Conductor-manufactured length refers to the length of the conductor measured at zero tension and at normal day temperature of 15–20°C. It is the same as the length of the conductor placed on the layers on reel.

6.4.2.6 Weight and mass of conductor

Normally weight or mass refers to the quantity with respect to 1 m. If weight is provided, mass is to be calculated by dividing weight by gravitational acceleration (9.81 m/s^2).

6.4.2.7 Annealing

Annealing is the process of reducing tensile strength at high temperature (>75°C).

6.4.2.8 Hardening of aluminium and aluminium alloy

Aluminium or aluminium alloy is used as stranded wire. Based on the hardening process, they are divided into the following two categories:

- non-heat treatable
- heat treatable

Different comparative features of hardening of non-heat treatable and heat treatable aluminium or aluminium alloy have been presented in Tables 6.5 and 6.6.

Table 6.5 Different features of hardening of non-heat treatable aluminium or aluminium alloy

Type	Hardening of non-heat treatable aluminium or aluminium alloy
Hardening and strengthening process	Plastic deformation Swaging Rolling Drawing
Strengthening	Proportional to deformation undergoes
Example	1350 Strands
Softening	Thermal treatment
Control of softening by	Temperature and time
O-temper	Refers to the state of fully tempered, also known as full annealing
Example	1350-O strands
Tensile strength	1350-O<1350
Ductility	1350-O>1350
Conductivity	1350-O>1350

Table 6.6 Different features of hardening of heat treatable aluminium or aluminium alloy

Type	Hardening of heat treatable aluminium or aluminium alloy
Hardening and strengthening process	Plastic deformation Swaging Rolling Drawing Thermal treatment Solution heat treatment (SHT) Ageing treatment (AT) Both of thermal treatment and plastic deformation
Strengthening	Proportional to deformation under goes
Example	Heat treatable aluminium alloy 6201
Materials	Aluminium with magnesium and silicon
Strength	6201>1350

6.4.2.9 Coating steel wires

Corrosion resistance of steel wires is improved by coating or galvanization. Based on thickness of coating or galvanization, steel wires are divided into Classes A, B and C. Thickness of coating for Class C is greater than Class B that is greater than Class A.

Sometimes aluminium is used as coating on steel. Thickness of aluminium coating (e.g. steel core used in ACSR/AZ) is less than the thickness of aluminium layer used in aluminium clad steel wire (e.g. steel core used in ACSR/AW).

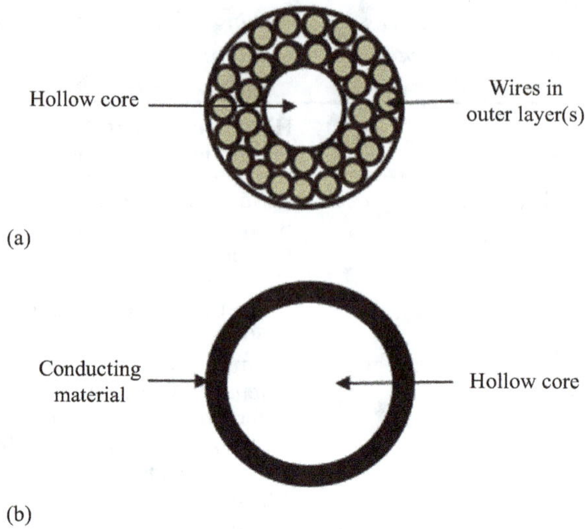

Figure 6.10 Hollow conductor (a) filled by wires of single materials, (b) made of sheets of single material

6.5 Hollow conductor

Hollow conductors are special type of conductors where core space is not filled by wire. It is kept either vacuum or filled by cooling and arc quenching medium. SF_6 or liquid nitrogen or their mixture is used as cooling medium. Outer layers may be made of sheets of single material or filled by wires of single materials as shown in Figure 6.10. However, use of hollow conductor is not normal seen in overhead line application. Very rare application is observed for around 100 m length in enclosed area.

6.6 Conductor with optical fibre cable

Conductor with optical fibre cable is special type of conductors where optical fibre cable is placed in core as shown in Figure 6.11. This cable is made of communication of high-frequency signals. Application of conductors with optical fibre cable is very limited on overhead lines but has future potential particularly in earth wire where load at normal condition is very low and can benefit communication for fault information as well as for other purpose. In protection scheme, this can be used as additional conductor to carry fault information. Optical ground wire is used for both rounding and communication purpose. It is also known as optical fibre composite ground wire. It finds application in earth wire in overhead lines.

6.7 Phase conductors

Phase conductors are the main part of overhead line conductors. The number of phase conductors depends on the type of overhead line whether it is distribution

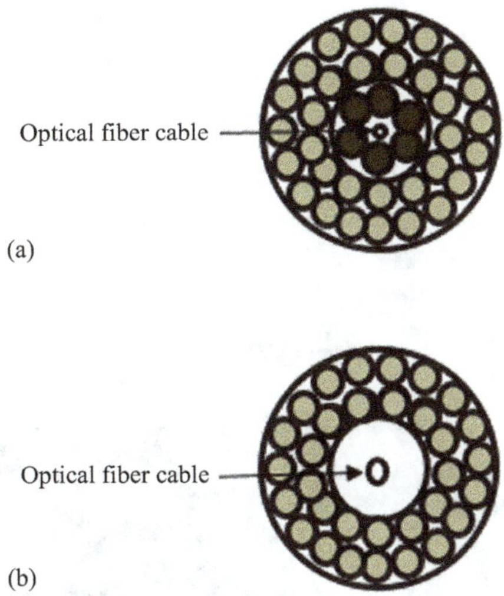

Figure 6.11 Hollow conductor with optical fibre cable (a) partly hollow core, (b) fully hollow core

line or transmission line. In low-voltage distribution line, phase conductors are accompanied by neutral ones. The number of phase conductors in low-voltage distribution line depends on the number of phases. For domestic distribution, one of three numbers of phases with one neutral is provided. In medium-high-voltage distribution lines, only phase conductors are present in three numbers without any neutral conductor. Transmission lines also run without neutral conductor.

The number of phase conductors in transmission and distribution lines from high to ultra-high-voltage range depends on the number of circuits carried by the lines (Figure 6.12). Total number of phase conductor becomes equal to the number of circuits (c) multiplied by three phases, i.e., $3c$. In the case of using bundled conductor as phase conductor, if the number of sub-conductors is b, then the total number of sub-conductors in the line will be $3bc$.

Phase conductors are installed on the pin-type insulator by suspension-/strain-type string insulators. Pin or string specifications are chosen on the basis of operating line voltage level. But conductors are chosen by the amount of electrical power or load to be carried out by the conductors. It should be less than current capacity or ampacity of the overhead line conductors.

6.8 Earth wire or sky wire

Earth conductor is installed for the overhead line, connected power system network and the safety of the people under the line. It is directly connected to the tower and grounded. Earth wire is placed above the phase conductor at the top of the tower. In this reason, it is also called sky wire (shown in Figure 6.13). The number of earth

Figure 6.12 Line conductors

Figure 6.13 Sky wire

conductor (1 or 2 or higher) depends on the voltage level of the overhead line and height of the tower. Continuous operating current of earth conductor is very less or negligible with respect to phase conductor but increases during faults.

6.9 Jumper

Jumpers refer to the conductors used to connect two ends of tension string. The same type of jumper material is normally used that is used for line conductors.

Jumpers are placed in three positions as found suitable for the support and insulator arrangements as follows:

1. Jumpers hanging through below (Figures 6.14 and 6.15)
2. Jumpers passing through top (Figure 6.16)
3. Jumpers passing by side of 'path of the line' (Figure 6.17)

Some real application of three types of jumpers in distribution system have been shown in Figure 6.18. Joining of jumper with insulator end and line conductor is very important and to be handled with skilled care failing which may result in failure.

6.10 Covered conductors or overhead cables

Covered conductor has the following major advantages in distribution system (Figure 6.19):

1. Probability of short circuit by wind effect reduces
2. Short circuit due to tree fall eliminates
3. Safety improves
4. Interference with other electrical or communication circuit reduces
5. Power theft reduces

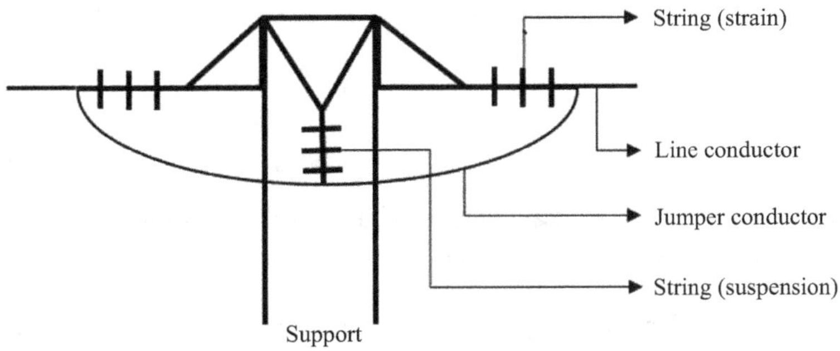

Figure 6.14 Jumper passing below the level of line conductor through string

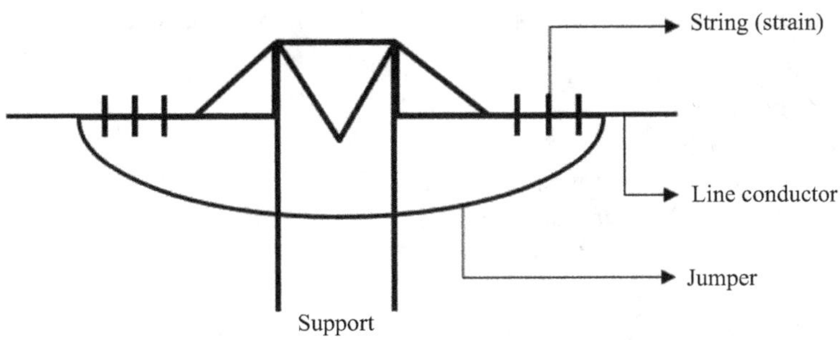

Figure 6.15 Jumper hanging below the line conductor

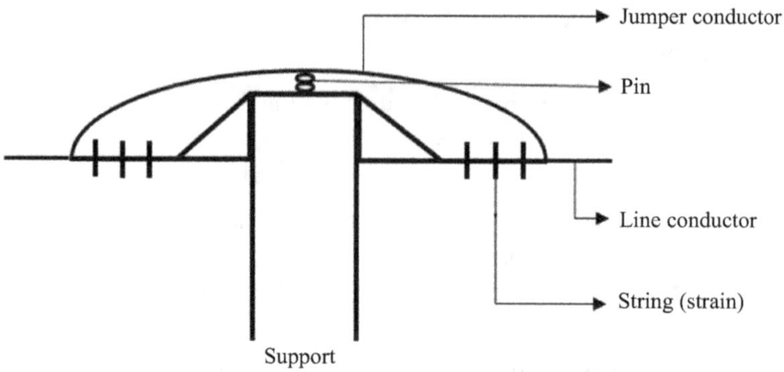

Figure 6.16 Jumper passing over the level of line conductor over insulator

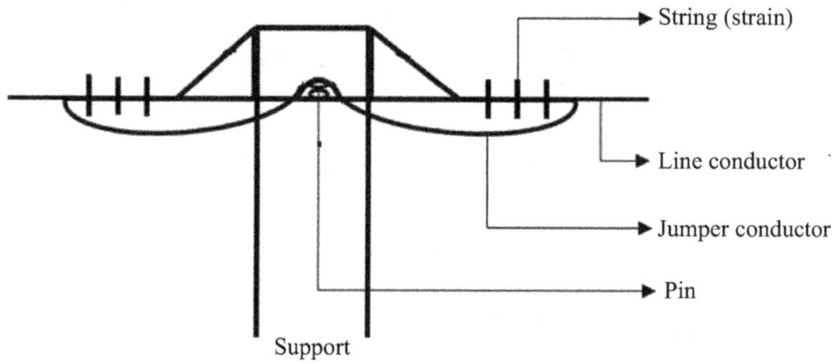

Figure 6.17 Jumper passing through the same level of line conductor over insulator

Covered conductors are used in overhead lines at low-voltage and medium-voltage distribution system. Covered conductors or cables consist of neutral and phase conductors. It may or may not consist of messenger wire. Conductors are covered by cross-linked polyethylene (XLPE) insulation. Stranded conductors made of aluminium or composite materials are used. Alloy made of Al, Mg, Si, etc. are used in messenger wire. The number of strands depends on gauge of the wire. Insulation level depends on the voltage level. Insulation must be tested at no-load and full-load conditions. Covered overhead conductors are specified by the following parameters:

- Operating voltage (kV)
- Conductor material
- Insulation
- Maximum operating temperature (°C)
- Maximum overload temperature (°C)
- Maximum short circuit temperature (°C)
- Cross-sectional area (mm^2)
- DC resistance at 20°C (Ω/km)
- AC resistance at 20°C (Ω/km)
- Current carrying capacity (A)

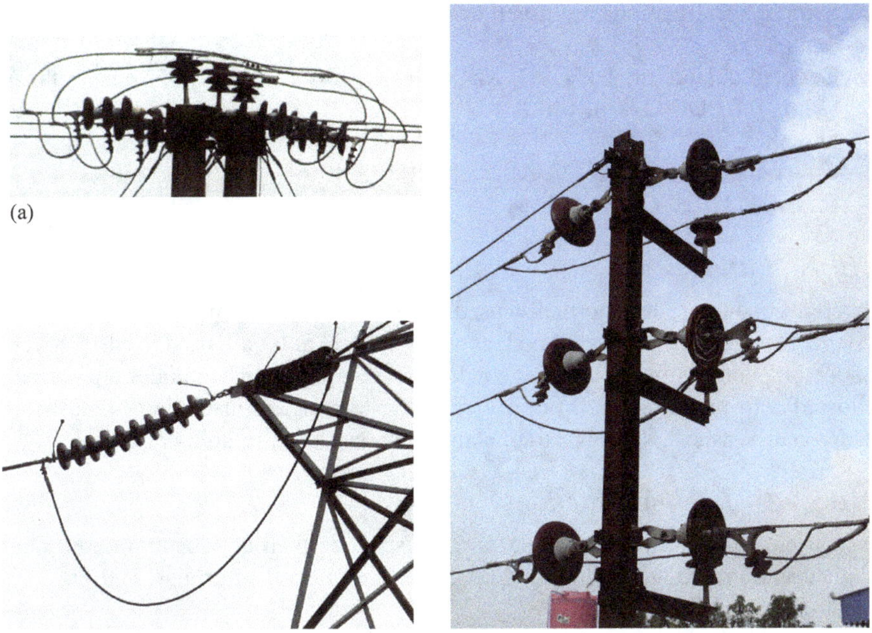

(a)

(b)

(c)

Figure 6.18 Jumper lines passing (a) through top, (b) below, (c) by side of the height of main line conductors

Figure 6.19 Distribution lines: from bare conductor to covered conductors

Sample specifications are given next:

Sample 1:

- Covered conductor, 1 kV, Al, XLPE, 90°C, 130°C, 250°C, 35 mm^2, 1.00 AC Ω/km, 0.87 DC Ω/km, 140 A

Sample 2:

- Covered conductor, 15 kV, Al, XLPE, 90°C, 130°C, 250°C, 35 mm^2, 1.00 AC Ω/km, 0.87 DC Ω/km, 140 A

6.10.1 Fittings

Covered conductors are normally used in low- and medium-voltage distribution system. Therefore, as support, wood, concrete or steel poles are used to carry covered conductors. For jointing of covered conductors, compression techniques are normally followed with pressure of 20 t or above. After jointing, re-insulation is must to be carried out. Washers, springs, strap, clamps, etc. are used to hold conductor.

6.10.2 Grounding practice

Neutral conductors and messenger wire must be earthed at substation end. Other than substation end, grounding is done at certain interval of around 300 m.

6.10.3 Tests

For covered conductors, routine tests as well as some specific tests are followed. For example, measurement of neutral to phase resistance, measurement of insulation resistance, short duration high-voltage withstanding test, partial discharge test, etc. are done as routine test. To see the ageing effect or effect of atmosphere, some special tests are carried out. In some cases, at pre-installation period, DC voltage withstanding test is carried out. Fault location identification test is carried out by creating artificial hole(s) at arbitrary points of insulation and testing whether it can be identified easily by sound, visually or by other means.

6.10.4 Cost comparison

Cost comparison [3] done among bare overhead conductors, covered overhead conductor and underground cables shows that cost involved in covered overhead conductor is approximately 1.3 times higher than that of bare overhead conductors, and cost of underground cables is approximately 4.3 times higher than that of covered overhead conductors. With respect to bare overhead conductor, cost of underground cables is much higher than that of covered overhead conductors. Therefore, covered overhead conductors may be good choice for safety and economics with respect to underground cables.

6.11 Current load

Current load is specified by current carrying capacity or ampacity (A). It is defined as the maximum amount of current that conductor can carry maintaining the security and safety of the conductor. It is also known as steady-state thermal rating or thermal rating of the line conductor.

6.12 Conductor fittings

6.12.1 Conductor on pin insulator

Conductors used with pin-type insulators are placed at the top and bounded by wires tightly. Conductor binding on pin-type insulator made of porcelain and composite material has been shown in Figure 6.20.

In practice, position of conductor on pin-type insulator is done in the following two ways:

- Conductors are positioned on the groove as shown in Figure 6.20(a). Care should be taken so that conductor is placed thoroughly on the groove and no-load concentration or slipping occurs during positioning. This followed when overhead line path follows straight line.
- Conductor positioned at side of the groove has been shown in Figure 6.20(b). It is followed when overhead line path bends, i.e., for angle support.

Commonly two types of binding methods are followed:

- Stirrup binding method
- Western union method

Binding wire or tape is used for positioning tightly. For aluminium-based conductors like aluminium alloy conductor (AAC) and aluminium conductor steel reinforce (ACSR), aluminium is used and for copper-based conductor soft copper is used as binding wire or tape materials.

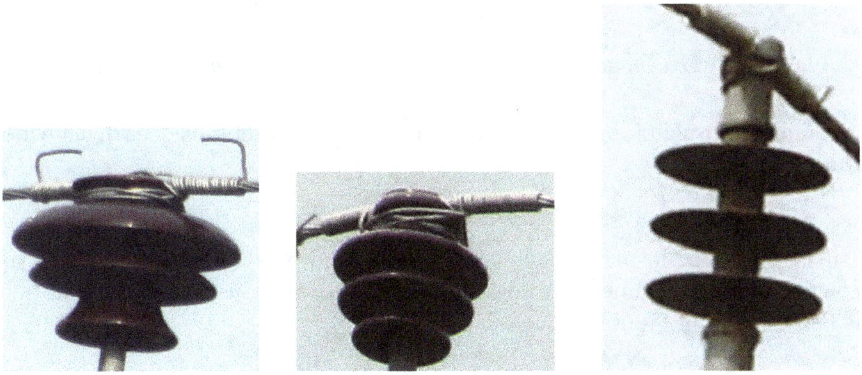

(a) (b) (c)

Figure 6.20 Conduct or positioning on different types of pin-type insulator (a) and (b) pin type porcelain insulators, (c) pin type composite insulator

Figure 6.21 Ceramic insulator end with lock for conductor fitting

Figure 6.22 Overhead line conductor fitted below suspension insulator disc

Ceramic insulator end with lock for conductor fitting has been shown in Figure 6.21.

6.12.2 Conductor with suspension-type disc insulator

In suspension-type insulators, conductors are placed by clamping below the string through clamp arrangement. Conductor placed below the suspension insulator disc has been shown in Figure 6.22.

6.12.3 Conductor with tension-type disc insulator

In tension-type insulator, conductors are connected with the end of string together with jumper by clamp arrangement. Conductor fitted with strain-type disc insulator has been shown in Figure 6.23.

Figure 6.23 Conductor fitted with strain-type disc insulator

6.12.4 Conductor with shackle-type insulator

Shackle-type insulators are held tightly by D- or U-shaped clamp arrangement. Conductor is positioned at the central part of the shackle-type insulator and bounded by tape of wire of preferably the same materials (Figure 6.24). This type of fittings is seen in low-voltage distribution lines.

6.12.5 Earth conductor with support

Earth conductors are connected at the top of metal tower directly and with the cross arms or support end of line insulators without using any insulator. Tight fittings of this earth conductor with cross arm confirm that practically no resistance is

Figure 6.24 Conductor fitted with shackle-type insulator

imposed in the path of this earth wire during fittings. Earth conductor ensures tower/support body to be at earth potential ensuring the safety of support.

6.12.6 Conductor spacing

Conductor spacing is an important area which should be taken care during installation as well as in routine check-up of the line. Space refers to the inter distance between line conductors. It is maintained by spacer. Spacers reduce line fault probabilities and unwanted discharges. Two types of spacers are used as follows:

1. Spacers for sub-conductors of bundled conductor
2. Spacers for phase conductor

The shape of spacer used for bundled conductor depends on the number of sub-conductors used in bundled form. Spacer used in 765 kV transmission line has been shown in Figure 6.25 for bundled conductor having six sub-conductors.

Example of spacer used in 440 V distribution line has been shown in Figure 6.26.

6.12.7 Reel

Reel is not direct component of overhead lines but an important item required for conductor storing and transportation. Reel used conductor in 765 kV line installation has been shown in Figure 6.27.

Reel has two parts: core and side guard. Core is made of either metal (aluminium, etc.) or wood. Metal core is reusable. Sides are made of either wood metal sheets of strong PVC.

6.12.8 Installation care

While installing conductor in overhead lines, the following care should be taken to reduce undesired conductor damages (Figure 6.28):

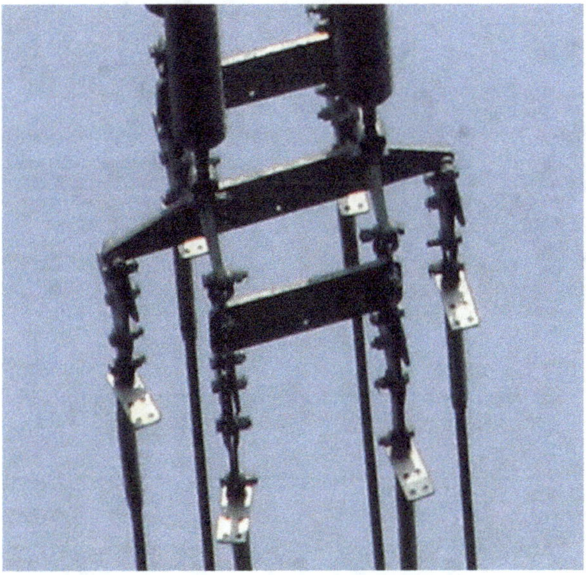

Figure 6.25 Spacer used in 765 kV transmission line

Figure 6.26 Spacer used for 440 V distribution lines

- Avoid dragging of surface
- Avoid re-winding of reel
- Avoid moving over rock surface
- Avoid scratching with fence or guard structure

(a) (b)

Figure 6.27 (a) Reel and (b) reel stand used for conductors in 765 kV lines

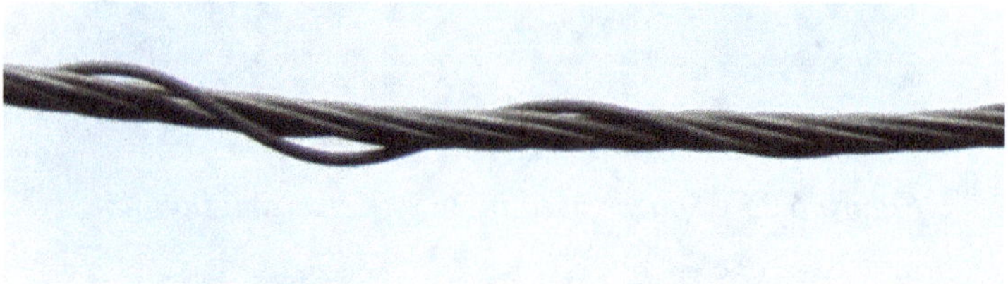

Figure 6.28 Damaged conductor

6.13 Common stringing method

Stringing methods commonly followed in conductor installation are as follows (Table 6.7):

- Tension method
- Semi-tension method
- Slack method
- Layout method

6.14 Tension methods

Tension methods used for tensioning conductor are divided into two major types as follows:

1. Tensioner with bullwheel
2. V-groove tensioner

Tensioner with bullwheel is again divided into the following types:

1. Offset multi-groove
2. Tilted multi-groove

Table 6.7 Different string methods and steps followed by them

Stringing methods	Major steps
Tension method	• Pulling line is set • Steel cable or synthetic rope is used for pulling • Puller wench is used at one end of conductor • Tensioner (bullwheel type) is used at other end • Reel is connected behind tensioner • Conductor is pulled through stringing sheave
Semi-tension method	• Pulling line is set • Steel cable or synthetic rope is used for pulling • Puller wench is used at one end of conductor • Puller directly pulls conductor from reel • Conductor is pulled through stringing sheave
Slack method	• Reel is placed on stand and made free to rotate • Vehicle pulls the conductor from other end • Stringing sheave or stringing block or traveller is used for stringing conductor at each structure
Layout method	• Reel is placed on vehicle or on trailer • Conductor is paced on sheave for stringing • Conductor is lifted • Conductor is pulled by the vehicle

3. Motorized bullwheel
4. Tensioner with bullwheel and payoff trailers

Groove dimension is very important part to be chosen as per the guidelines of conductor manufacturer. IEEE 524 standard may be referred to in this regard.

For ACSR, AAC, ACAR, ACSR/TW, all aluminium alloy conductor and AACSR conductor of diameter (D_c), the following values may be considered as limiting lower limit.

1. Radius of groove: 55% of conductor diameter
2. Groove depth: 25% greater than conductor diameter
3. Groove flare angle: between 15° and 20°

If two conductors are placed, then magnitude of D_c is to be considered as twice the diameter of the one conductor. Groove dimension and angles should be carefully chosen; too flat or too narrow groove can damage conductor considerably.

6.14.1 Machine for stringing

After foundation and installation work of overhead lines, stringing is carried out. This machine (shown in Figure 6.29) helps in adding and adjusting conductor tension, sag, etc. to a proper value.

Figure 6.29 Machine used for stringing conductors in 765 kV lines

6.15 Design features

Design features [3] of overhead line conductors consider as follows*:

- Electrical properties like resistance, inductance and their effects
- Mechanical dimension and sags and their effect
- Thermal factors

Different design features have been discussed in following subsections.

6.15.1 DC resistance

As in composite conductors, different materials are used in core and outer layers. DC resistance of the conductor is contributed by DC resistances of both core and outer layers.

6.15.1.1 Resistance of steel core

Let z_{st} be the number of steel wires, d_{st} be the diameter of steel wires, ρ_{st} be the resistivity of steel wires, K_{st} be the constant for lay length factor of steel wires, n_{st} be the number of steel wires, and R_{st} be the resistance of steel core.

*https://elek.com.au/articles/factors-affecting-bare-conductor-current-ratings/.

Therefore, the resistance of steel core (R_{st}) can be written as follows:

$$R_{st} = \frac{\rho_{st}}{\frac{\pi d_{st}^2}{4}} \frac{1}{\left(1 + \sum_1^{n_{st}} \frac{6z_{st}}{K_{st}}\right)} = \frac{4\rho_{st}}{\pi d_{st}^2 \left(1 + \sum_1^{n_{st}} \frac{6z_{st}}{K_{st}}\right)} \tag{6.1}$$

6.15.1.2 Resistance of outer layer of aluminium wires

Let z_{Al} be the number of aluminium wires, d_{Al} be the diameter of aluminium wires, ρ_{Al} be the resistivity of aluminium wires, K_{Al} be the constant for lay length factor of aluminium wires, n_{Al} be the number of aluminium wires and R_{Al} be the resistance of outer layer of aluminium wires.

Therefore, resistance of steel core can be written as follows:

$$R_{Al} = \frac{\rho_{Al}}{\frac{\pi d_{Al}^2}{4}} \frac{1}{\sum_1^{n_{Al}} \frac{6z_{Al}}{K_{Al}}} = \frac{4\rho_{Al}}{\pi d_{Al}^2 \sum_1^{n_{Al}} \frac{6z_{Al}}{K_{Al}}} \tag{6.2}$$

Total number (n) of wires is

$$n = n_{st} + n_{Al}$$

Electrically, steel cores and aluminium wires are in parallel. Therefore, DC resistance of the conductor (R_{DC}) may be written as follows:

$$\frac{1}{R_{DC}} = \frac{1}{R_{Al}} + \frac{1}{R_{st}}$$

$$\text{Therefore, } R_{DC} = \frac{R_{Al} R_{st}}{R_{Al} + R_{st}} \tag{6.3}$$

6.15.1.3 Temperature coefficient of resistance

Let T_1 be the reference temperature in °C, T_2 be the desired temperature in °C, T be the temperature difference in °C, R_1 be the resistance at T_1°C, R_2 be the resistance at T_2°C and α_T be the temperature coefficient resistance.

Resistance at T_2°C may be written as follows:

$$R_2 = R_1\{1 + \alpha_T(T_2 - T_1)\} = R_1(1 + \alpha_T T) \tag{6.4}$$

Therefore,

$$\alpha_T = \frac{R_2 - R_1}{T} \tag{6.5}$$

6.15.2 Inductance

Inductance of a transmission line conductor forms due to the flux linkage which occurs both internally and externally. Thus, total inductance of a conductor can be mathematically expressed as follows:

$$L_T = L_{in} + L_{ex} \tag{6.6}$$

where L_{in} is the inductance occurred due to internal flux linkage and L_{ex} is the inductance occurred due to external flux linkage.

Internal inductance is given as follows:

$$\begin{aligned} L_{in} &= \frac{\mu}{8\pi} \\ &= \frac{1}{2} \times 10^{-7} \mathrm{H/m} \end{aligned} \tag{6.7}$$

External inductance power loss is given as follows:

$$\begin{aligned} L_{ex} &= \frac{\varphi_{ex}}{I} \\ &= \frac{4\pi \times 10^{-7}}{2\pi} \ln \frac{D}{r} \\ &= 2 \times 10^{-7} \ln \frac{D}{r} \ \mathrm{H/m} \end{aligned}$$

Total inductance may be written as follows:

$$\begin{aligned} L_T &= \frac{1}{2} \times 10^{-7} + 2 \times 10^{-7} \ln \frac{D}{r} \\ &= 2 \times 10^{-7} \left(\frac{1}{4} + \ln \frac{D}{r} \right) \\ &= 2 \times 10^{-7} \ln \frac{D}{re^{-1/4}} \\ &= 2 \times 10^{-7} \ln \frac{D}{r'} \end{aligned} \tag{6.8}$$

where

$$r' = re^{-1/4} \tag{6.9}$$

Thus, total inductance of a conductor of radius r is equal to the inductance due to external flux linkage of a conductor of radius $r' \left(= re^{-1/4}\right)$

In terms of geometrical mean distance (GMD) and geometrical mean radius (GMR), inductance of a line can be written as follows:

$$\overline{L_T} = 2 \times 10^{-7} \ln \frac{GMD}{GMR} \tag{6.10}$$

6.15.3 Skin effect

Non-uniformity of resistance is almost the same in both AC and DC. But in AC, other than non-uniform resistive effect, there is effect of non-uniform flux linkage. Internal flux linkage is high at the centre and minimum at the periphery. Thus, overall impedance is high at centre and minimum at periphery. However, skin effect is more in AC than in DC.

Thus, skin effect can be defined as a phenomenon by which current density becomes high near to the periphery and becomes low near to the centre of the conductor.

Conductor loss considering skin effect is greater than loss without skin effect. Thus, skin effect increases effective resistance and hence net loss.

6.15.4 Proximity effect

When two or more conductors are placed nearby and current flow through them, attraction or repulsion takes place on the basis of the direction of current flow whether they are in the same or in the opposite direction. This results in unequal distribution of current density. This phenomenon is known as proximity effect. This increases effective resistance of the conductors and loss components.

6.15.5 Effective AC resistance

AC resistance of conductors is greater than their DC resistance value. Skin effect and proximity effect increase effective AC resistance of the conductor. Effective AC resistance may be determined by power loss divided by square of the current flowing through the conductor.

6.15.6 Sags

Mechanical sags of overhead line conductors have been discussed in Chapter 2. Sag (y_{SAG}) of the conductor at midpoint of the span of the conductor over a length of $2L$ may be written as follows:

$$y_{SAG} \cong \frac{wL^2}{8H} \tag{6.11}$$

where w is the weight of the conductor per unit length, L is length between two supports, and H is horizontal pressure.

Contributors to the sag may be divided into two categories as follows:

1. Physio-mechanical
2. Electro-thermal

Sags considering various factors have been shown in Figure 6.30.

6.15.6.1 Physio-mechanical contribution

Parameters for sag calculation are considered at 15°C. Sag calculated considering weight, horizontal tension and span of the line may be considered as the initial sag which is experienced at initial stage of installation.

Let,

y_{SAG-I} = sag at initial installation

Now, the conductor for which sag is considered is joined with the end of insulator or string of insulators which are subjected to either suspension or tension.

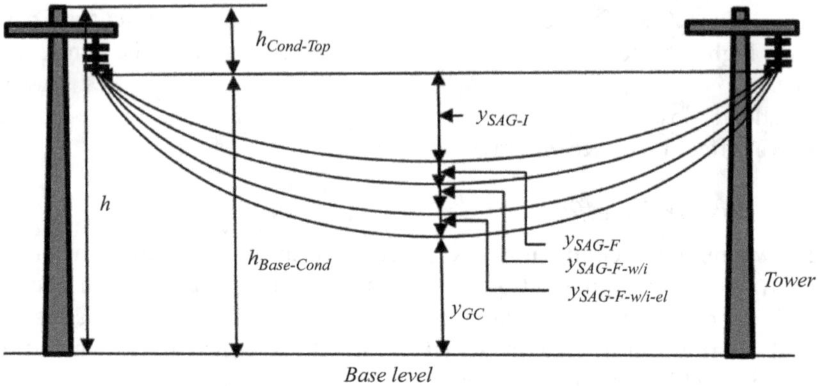

Figure 6.30 Sags considering various factors

Moreover, they also undergo different longitudinal and transverse oscillation. All these features will contribute to the sag.

Let,

y_{SAG-F} = sag at the final stage considering the effect of insulator arrangement.

Therefore,

$$y_{SAG-F} > y_{SAG-I}$$

In addition to the earlier, there will be effect or wind or ice or both.

Let,

$y_{SAG-F-w/i}$ be the sag adding the effect of wind or ice or both to the earlier (y_{SAG-F}).

Therefore,

$$y_{SAG-F-w/i} > y_{SAG-F} > y_{SAG-I}$$

6.15.6.2 Electrical contribution

Factors considered earlier are not electrical. However, conductors are installed to carry electrical loads on the basis of the current carrying capacity or ampacity. Depending upon the capacity, it carries load. Based on electrical parameters and connected devices, current carried by the conductor varies at different load and fault conditions. Thermal behaviour of the conductor is reflected on the basis of the temperature status of the conductor resulted from various electrical load. Thermal behaviour of the conductor materials at high temperature contributes to the load. Let $y_{SAG-F-w/i-el}$ be the sag adding the effect of electrical loads to the $y_{SAG-F-w/i}$.

Therefore,

$$y_{SAG-F-w/i-el} > y_{SAG-F-w/i} > y_{SAG-F} > y_{SAG-I}$$

6.15.7 Ground clearance

Sufficient ground clearance below the conductor of overhead lines is required for safety for all beings having access at the base level below the lines. Therefore, starting from the design it must be considered. Ground clearance y_{GC} considering sag (y_{SAG}) and height (h) of the tower may be written as follows:

$$\text{Ground clearance} = y_{GC} = h - \left(y_{SAG} + h_{Cond-Top}\right) \tag{6.12}$$

where h is the height of the line support, $h_{Cond-Top}$ is the distance between lower conductor joint point with line insulator and the top of the tower.

$$\text{Ground clearance} = y_{GC} = \left(h - h_{Cond-Top}\right) - y_{SAG}$$

$$\text{Ground clearance} = y_{GC} = h_{Base-Cond} - y_{SAG}$$

where $h_{Base-Cond}$ is the height of the lower most conductor at tower from base level.

$$y_{SAG} = y_{SAG-F-w/i-el}$$

Therefore, ground clearance may be written as follows:

$$\text{Ground clearance} = y_{GC} = h_{Cond} - y_{SAG-F-w/i-el} \tag{6.13}$$

For the safety of all beings having access at the base level below the lines, ground clearance must be greater than the specified value as provided by the regulating authority of the region. It depends on the following:

- Voltage level
- Tower height
- Distance from top to lower conductor position on the tower
- Locality
- Whether road, river is there under the line
- Whether there is any construction under the line. If present, their height located
- Whether the line is crossing rail
- Whether the line is crossing over communication line

6.15.7.1 Volume, weight and tension

Volume of conductor is estimated by multiplying area and length of the conductor as follows:

$$v = \text{Area} \times \text{length}$$

While estimating cross-sectional area, it is assumed as perfectly circle (it does not introduce much errors as diameter is much less than length of the conductor).

Area may be written as follows:

$$\text{Area} = \frac{\pi d^2}{4}$$

Therefore, volume may be written as follows:

$$v = \frac{\pi d^2}{4} \times \text{length} \tag{6.14}$$

Let average mass density be ρ.
Then, total mass (M) may be written as follows:

$$M = \rho \times \frac{\pi d^2}{4} \times \text{length} \tag{6.15}$$

Total weight (W) may be written as follows:

$$W = Mg = \rho \times \frac{\pi d^2}{4} \times \text{length} \times g \tag{6.16}$$

Weight per unit length (w) may be written as follows:

$$w = \frac{W}{\text{length}} = \frac{Mg}{\text{length}} = \rho \times \frac{\pi d^2}{4} \times g \tag{6.17}$$

This weight contributes to the tension–stress relation of the conductor.

6.15.8 Stringing chart

Both sag and tensile strength vary with temperature. The relation of sag and tensile strength with temperature considering ice and wind effect is known as stringing chart. Values for sag and tensile strength are their limiting values corresponding to specific temperature to be considered for safe operation of the conductor. With the increase of temperature, sag increases, whereas tensile strength decreases. Typical stringing chart has been shown in Figure 6.31.

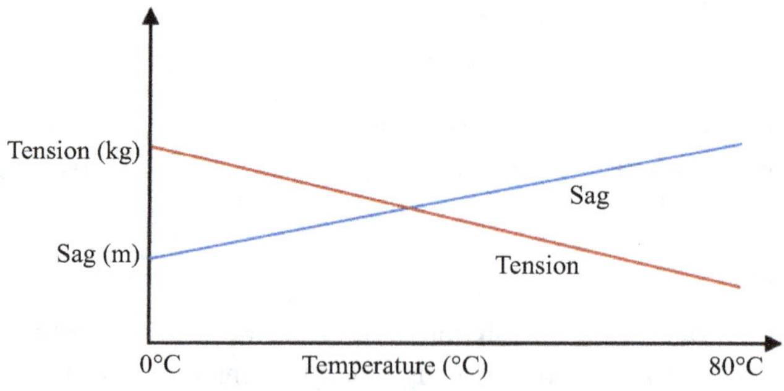

Figure 6.31 Stringing chart

6.15.9 Percentage slack

Slack is a physical dimensional property of overhead line conductor. It is defined as the ratio of difference between conductor length and span to the span. Mathematically, it may be written as follows:

$$\text{Slack}(S_L) = \frac{\text{Length of conductor} - \text{Span}}{\text{Span}} \tag{6.18}$$

Percentage of slack may be written as follows:

$$\text{Percentage slack} = \%S_L = \frac{\text{Length of conductor} - \text{Span}}{\text{Span}} \times 100\% \tag{6.19}$$

6.15.10 Slack–stress relation

Stress on the conductor is related to slack. Relation has been shown in Figure 6.32 with the increase of percentage of slack stress decreases exponentially.

6.15.11 Slack–sag relation

Sag on the conductor is related to slack. Relation has been shown in Figure 6.33. With the increase of percentage of slack, sag increases exponentially.

6.15.12 Selection of conductor

Selection of conductors is done by correlating the requirement of the overhead line, conductor availability as provided by the manufacturer, structure and cost. Like other devices, transmission and distribution lines are not built with expectation of short period of life. A line is expected to run for a long period of time. Conductor is the main part of line. Accordingly, other parts of the line like support, insulators and protective systems are designed. In other words, conductor can be considered as nucleus of the whole transmission and distribution family. Therefore, the selection of conductor is very important for designers. Following major factors are considered for the selection of conductors:

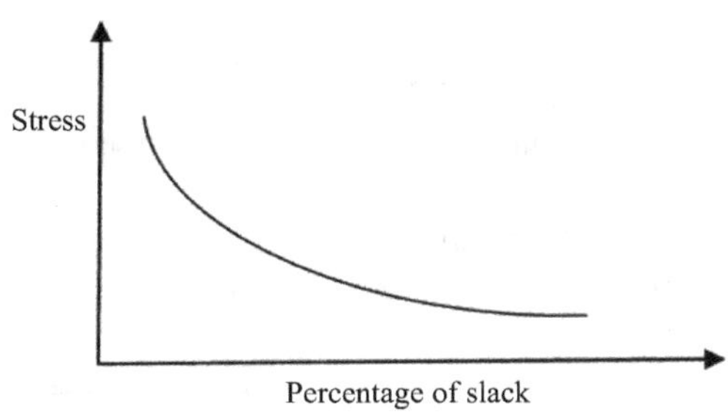

Figure 6.32 Stress versus slack relation

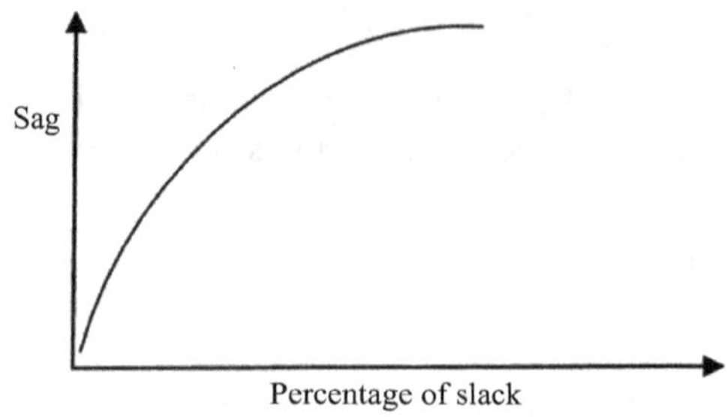

Figure 6.33 Stress versus slack relation

1. Desired current capacity
2. Permissible and possible current carrying capacity or ampacity
3. Limit of tensile strength
4. Operating tensile strength
5. Sag
6. Sag in the worst condition
7. Electrical losses
8. Corona and other noise
9. Interference with communication lines
10. Operating temperature
11. Maximum allowable conductor temperature
12. Type and dimension of the conductor
13. Cost

6.16 Conductor temperature

6.16.1 Temperature variation

Electrical power transmitted through line conductor produces heat that is proportional to the current flowing through it. Magnitude of current depends on power flow and voltage rating of the line. Heat increases the temperature. Temperature of conductor depends first on square of the current or ampere value of the conductor. Along with electrical conduction, heat conduction takes place in the conductor that depends on thermal conductivity of the conducting materials. Thermal conductivity of copper is much higher than aluminium. Thermal conductivity of steel is much less. Thermal conductivity of air is small. But heat conduction in air takes place by convection and radiation.

Considering all these, temperature of a conductor is not uniform over the cross-sectional area. Temperature variation with current flow has been shown in Figure 6.34. Temperature variation along the radius has been shown in Figure 6.35. It shows that temperature is high at centre and then reduces. Rate of temperature reduction is slow near central area. Then temperature decreases at higher rate near peripheral area and reaches surface temperature. Surface temperature depends on heat dissipation from the surface area.

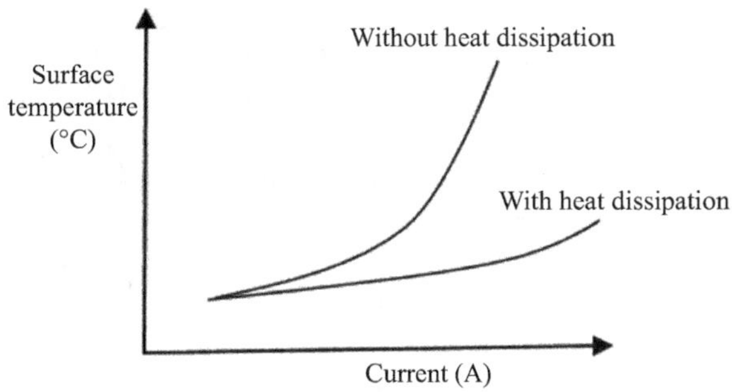

Figure 6.34 Surface temperature versus current

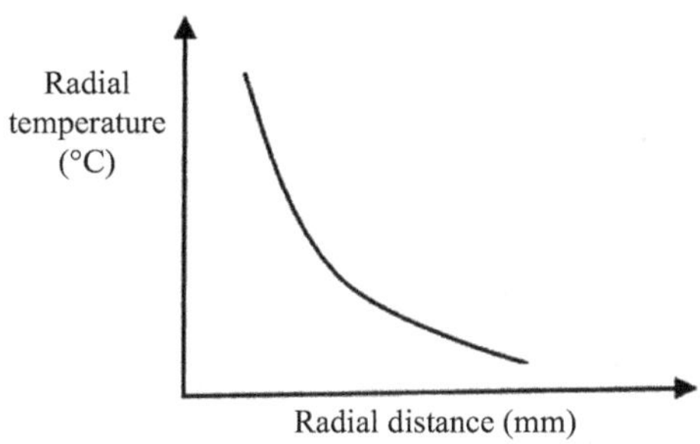

Figure 6.35 Temperature versus radial distance from centre

Heat dissipation increases with the increase of air flow. Variation of surface temperature with respect to wind speed has been shown in Figure 6.36(a). It shows that temperature at surface reduces almost exponentially with the increase of wind speed. Direction of wind velocity with respect to conductor is an important consideration. At high wind velocity normal to conductor, temperature becomes low at surface area. In other words, to keep surface temperature constant at high current, greater wind speed is required (Figure 6.36(b)).

6.16.2 Heat balance for conductor

Heat balance for overhead line conductor is governed by heat gain and heat loss. For overhead lines, there are two major components of heat gain: current and solar radiation.

Let,

Heat gain due to current be Q_A (W/km)

Heat gain from solar radiation be Q_S (W/km)

Heat loss may occur due to mainly two reasons as follows: heat convection and heat radiation.

Let,

Heat loss due to convection be Q_C (W/km)

Heat loss due to radiation be Q_R (W/km)

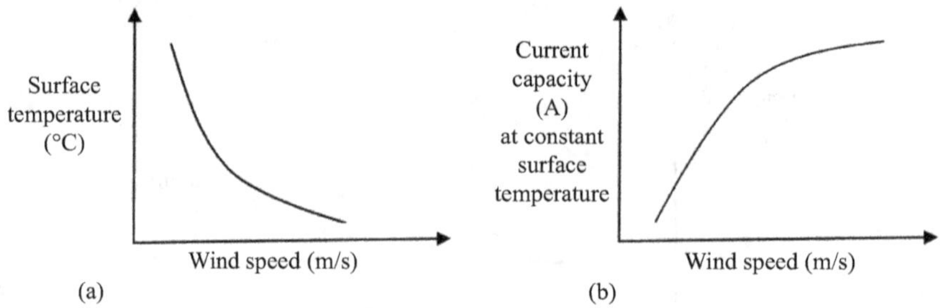

Figure 6.36 (a) Wind speed versus surface temperature versus (b) Wind speed versus current capacity at constant surface temperature

Therefore,

$$\text{Total heat gain} = Q_A + Q_S$$

$$\text{Total heat loss} = Q_C + Q_R$$

For operation at thermal balance condition,

$$\text{Total heat gain} = \text{total heat loss}$$

$$Q_A + Q_S = Q_C + Q_R$$

Heat gain due to current may be obtained from

$$Q_A = I^2 R$$

where resistance is

$$R = R_{20}\{1 + \alpha(\text{ambient temperature} + \text{temperature rise})\}$$

$$R_{20} = \text{resistance at } 20°\text{C}$$

$$\alpha = \text{temperature coefficient of resistance}$$

Heat gain due to solar radiation may be obtained from

$$Q_S = \text{solar absorption coefficient} \times \text{solar radiation intensity} \times \text{conductor diameter}$$

Solar absorption coefficient is less than one and it varies normally from 0.4 to 0.9. For bright and new conductor, its value is small and increases when conductor becomes old and black or less bright. Heat loss due to convection may be obtained from the empirical relation:

$$Q_C \propto (\text{wind velocity perpendicular to conductor} \times \text{conductor diameter})^i \times \text{temperature rise}$$

Magnitude of i is approximately 0.4, and proportionality constant is taken as 380–400. Heat loss due to radiation may be obtained from the empirical relation:

$$Q_R \propto \text{Thermal emisivity of conductor} \times \text{conductor diameter}$$

Proportionality constant is taken as $\pi \times$ Stefan Boltzmann's constant, where the constant is equal to $5.7 \times 10^{-8} \mathrm{W/m^2}$.

6.16.3 Temperature-dependent conductor-type selection

Temperature is an important physio-thermal parameter which is related to electrical loads as well as atmospheric temperature. It also relates mechanical behaviour of the conductor like sag and tensile strength. Therefore, temperature is one of the important guiding factors for selecting conductor type. Based on operating maximum temperature, conductors are categorized into two types as follows:

- Conductor of maximum temperature <100°C
- Conductor of maximum temperature >100°C

Different conductor types belonging to the previous categories have been presented in Table 6.8.

6.17 Conductor vibration

6.17.1 Classification of conductor motion

Conductors connected in overhead lines are exposed to climatic situation as well as different electrical conditions. Due to various reasons, overhead line conductors

Table 6.8 Operating maximum temperature-based conductors' types

Conductor of maximum temperature <100°C (low temperature: LT)	**Conductor of maximum temperature >100°C (high temperature: HT)**
All aluminium alloy conductor (AAAC)	TACSR (or ZTACSR): aluminium conductor steel reinforced (galvanized) consists of zirconium aluminium of thermal resistant of type AT1 or AT3
Aluminium conductor alloy reinforced (ACAR)	GTACSR (or GZTACSR): consisting of gap between core and aluminium layers
Trapezoid-shaped-wire conductors (AAAC/TW, ACSR/TW)	TACIR (or CTACIR): core is made of galvanized or aluminium clad invar alloy steel; outer layers consist of (Z)TAL
Motion-resistant conductor (self-damping conductor – SDC or conductor with twisted pair TP)	ACCR ceramic-fibre-reinforced core surrounded by AT3 zirconium aluminium alloy strands
	ACCC: core is made of single strand covered with carbon fibre (very low thermal elongation, reinforced in high-temperature thermoset resin), core is surrounded by outer layers of trapezoid annealed aluminium (TW)

undergo different types of motion. Conductor motion is mainly divided into two categories as follows:

1. Sustained wind-driven motion
2. Non-sustainable motion

Sustained wind-driven motion is again may be divided into three main categories as follows:

1. Aeolian vibration
2. Sub-span vibration or wake wind oscillation
3. Galloping

Non-sustainable motion includes the following categories of motion:

1. Motion due to power line short circuit forces
2. Oscillation due to corona
3. Oscillation due to ice accretion

All these conductor motions or vibration have adverse effect to the overhead line conductors and the other mechanical structure connected with conductors. This may result in deformation of the stranded layers to severe damage of disconnection of the line and breakage of the structure. Conductors are exposed to wind-generated forces which vary with variation of wind nature. Therefore, these vibrations should not be neglected from the beginning of foundation and installation; otherwise, it may lead to mechanical instability, power insecurity and economic loss of the system. Wind-driven motion can be classified into three categories with respect to relative frequency as follows:

1. High frequency in the range of 5–100 Hz: aeolian vibration belongs to this category. Amplitude of this vibration may reach to the height equal to half of the diameter of the conductor.
2. Low frequency in the range of 1–5 Hz: wake wind oscillation belongs to this category. Amplitude of the vibration may be equal to half of the distance between sub-conductors of bundled conductor.
3. Very low frequency in the range less than 0.5 Hz: galloping belongs to this category. Amplitude of this vibration may be as high as equal to the sag within the free span.

6.17.2 Aeolian vibration

6.17.2.1 Nature and cause

Wind-driven vibration of vertices of conductor from top to bottom is known as aeolian vibration. Wind of low speed in the range of 0.5–7 m/s causes this vibration. Von Karman first gave an explanation of generation of aeolian vibration. This is known as Von Karman effect. Due to wind speed, detachment vortices take place. Movement of vortices creates unbalanced pressure distribution. Unbalanced pressure distribution by vortices occurs due to wind blow generated force to oscillate conductors. Ups and downs oscillation of conductor takes place at right angle to the direction of wind flow.

According to Strouhal's formula, frequency of aeolian vibration (f) is proportional to wind speed (v_w m/s). It may be written as follows:

$$f \propto v_w \tag{6.20}$$

or,

$$f = \frac{d}{S} v_w \tag{6.21}$$

where d is diameter of the conductor in m, s is called Strouhal number. Its magnitude is slightly lower than Reynolds number. Its average value is taken 0.185.

Tension T (N) of the conductor has relation with this frequency as follows:

$$T = f^2 \lambda^2 m \tag{6.22}$$

where m is mass of conductor per metre length, λ is the wavelength of vibration (m).

$$f^2 = \frac{T}{\lambda^2 m} \tag{6.23}$$

$$f = \frac{1}{\lambda} \sqrt{\frac{T}{m}} \tag{6.24}$$

$$\frac{d}{S} v_w = \frac{1}{\lambda} \sqrt{\frac{T}{m}} \tag{6.25}$$

$$\sqrt{\frac{T}{m}} = \frac{d\lambda}{S} v_w \tag{6.26}$$

$$\frac{T}{m} = \left(\frac{d\lambda v_w}{S} \right)^2 \tag{6.27}$$

$$T = m \left(\frac{d\lambda v_w}{S} \right)^2 \tag{6.28}$$

6.17.2.2 Free span vibration angle

Free span vibration angle refers to the maximum angle created by the amplitude of vibration. Let A_{max} be the maximum amplitude of aeolian oscillation.

Therefore, free span vibration angle may be written as follows:

$$\gamma = \frac{A_{max}}{\frac{\lambda}{2\pi}} = \frac{2\pi A_{max}}{\lambda} \tag{6.29}$$

Now, frequency is

$$f = \frac{1}{\lambda}\sqrt{\frac{T}{m}}$$

or,

$$\lambda = \frac{1}{f}\sqrt{\frac{T}{m}}$$

$$\frac{1}{\lambda} = f\sqrt{\frac{m}{T}} \tag{6.30}$$

$$\gamma = \frac{2\pi A_{\max}}{\lambda} = 2\pi A_{\max} f\sqrt{\frac{m}{T}} \tag{6.31}$$

6.17.2.3 Effect

Aeolian vibration has many adverse effects; it includes dislocation of stranded wire, birdcage formation in stranding, increase of corrosion, etc. It also contributes change in conductor tension applied to insulator string.

6.17.2.4 Remedy

Damper and spacer dampers are used to reduce sub-span or wake wind vibration. Number and shape of the dampers and spacers vary line to line and probable statistics of wind in the area.

6.17.3 Wake wind oscillation

6.17.3.1 Nature and cause

Wake wind vibration is seen to have frequency in the range of 1–5 Hz. Wind of speed between 5 m/s and 6 m/s causes this type of vibration. Amplitude can reach up to half of the distance between sub-conductors of bundled conductor. It is also known as sub-span vibration. Mainly wake wind vibration of the following natures are observed:

- Horizontal
- Vertical
- Galloping

Frequency of sub-span vibration may be written as follows:

$$f = \frac{n}{\lambda l}\sqrt{\frac{T}{m}} \tag{6.32}$$

where n is constant which depends on the nature of the sub-span. Its value is considered as 1 or 2 and l is length of subspan.

The relation shows that frequency of sub-span oscillation varies.

- Proportional to the square root of tension
- Inversely proportional to the mass of the conductor
- Inversely proportional to the length of the sub-span

6.17.3.2 Effect

It has many adverse effects like dislocation of stranding wires, corrosion, etc. It may lead to intertouching of sub-conductors used in bundled conductor.

6.17.3.3 Remedy

Damper and spacer dampers are used to reduce sub-span or wake wind vibration.

6.17.4 Galloping

6.17.4.1 Nature and cause

Galloping is not observed frequently and is rare. It occurs due to sudden large change in aerodynamic behaviour of the conductor. It is observed in individual phase conductor as well as in bundled conductor. Its frequency is very small but amplitude can reach up to half of the sag. Asymmetrical distribution of ice on conductor and its interaction with wind may change the severity of galloping considerably.

6.17.4.2 Effect

Galloping may cause intertouching between phase conductor causing short circuit fault for very short period of time.

6.17.4.3 Remedy

Probability and severity of conductor galloping can be adjusted by proper design of sag, span of conductor with respect to height of the tower.

6.17.5 Damper

Dampers are used to reduce conductor vibration. They are connected near insulator end and sometimes at away from ends. Damper, as shown in Figure 6.37, has three parts: clamp, messenger and mass unit. It holds conductor by clamp. Clamp is connected to the mass unit by messenger. Mass units are located at two sides and used for reducing various types of vibration. The number of dampers is made either one or more than one. If the number of dampers is made more than one, it is made

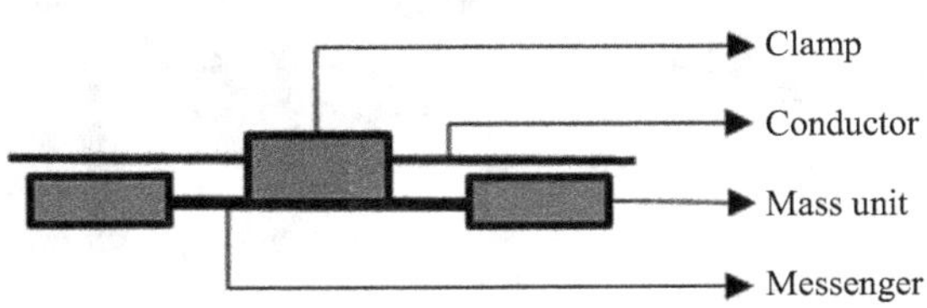

Figure 6.37 Different parts of damper

in even number and distributed equally from two supports of the line. Damper numbers are presented in Table 6.9.

Dampers connected in two sides of conductor have been shown in Figure 6.38.

Dampers connected in conductor used in 132 kV line has been shown in Figure 6.39.

Table 6.9 Damper numbers

Number of damper(s) near support 1	Number of damper(s) near support 1	Number of damper(s)
1	0	1
0	1	1
1	1	2
2	2	4
3	3	6

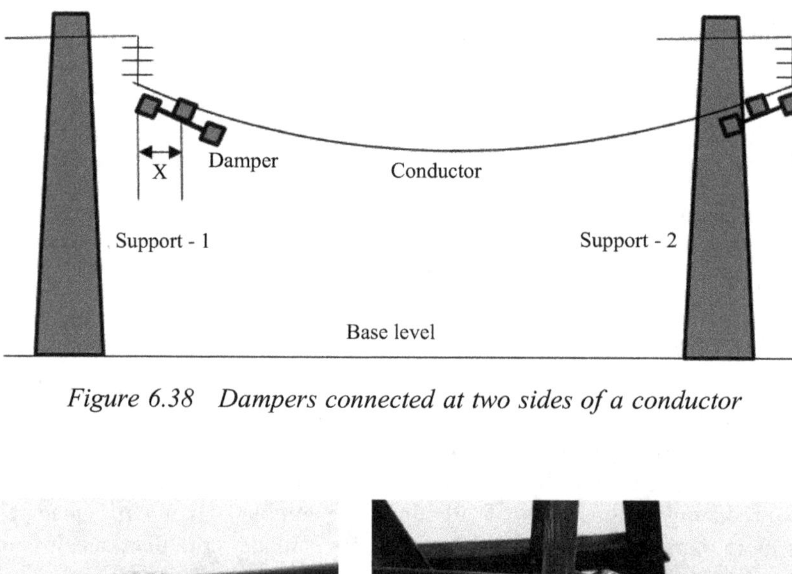

Figure 6.38 Dampers connected at two sides of a conductor

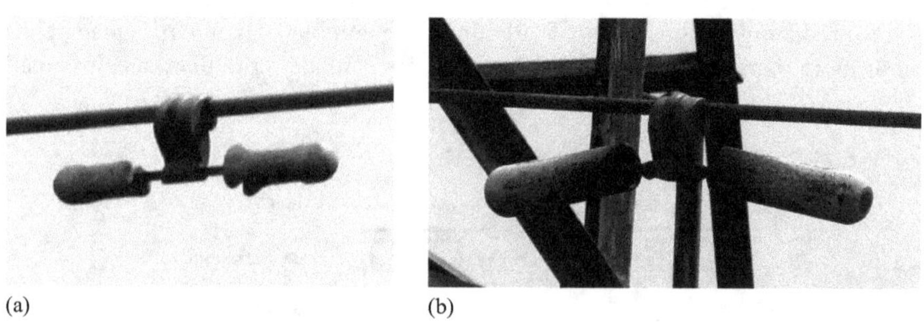

Figure 6.39 Damper in 132 kV line: (a) aged damper and (b) damaged damper

6.17.5.1 Distance of damper from support end

Distance of damper from the support end (x) is normally taken about 70–80% of half of the wavelength of oscillation. Now, wavelength may be written as follows:

$$\lambda = \frac{1}{f}\sqrt{\frac{T}{m}}$$

Therefore,

$$x \cong K_{DF}\frac{\lambda}{2} \tag{6.33}$$

where K_{DF} is the multiplying factor used for deciding the position of the damper from end with considering wavelength of oscillation. Its value is 0.7–0.8.

$$x \cong K_{DF}\frac{\lambda}{2} = K_{DF}\frac{1}{2f}\sqrt{\frac{T}{m}} \tag{6.34}$$

Frequency can be obtained from

$$f = \frac{d}{S}v_w \tag{6.35}$$

where d is diameter of the conductor in m, s is Strouhal number and v_w is wind velocity.

6.17.5.2 Spacer damper

Spacers are used in bundled conductor. Based on the number of sub-conductors used in bundled conductor, shape of the spacer is designed. By using mass components of the spacer, it is also used as damper, called spacer damper to reduce different types of conductor vibration. Different shapes of spacer dampers have been shown in Figure 6.40.

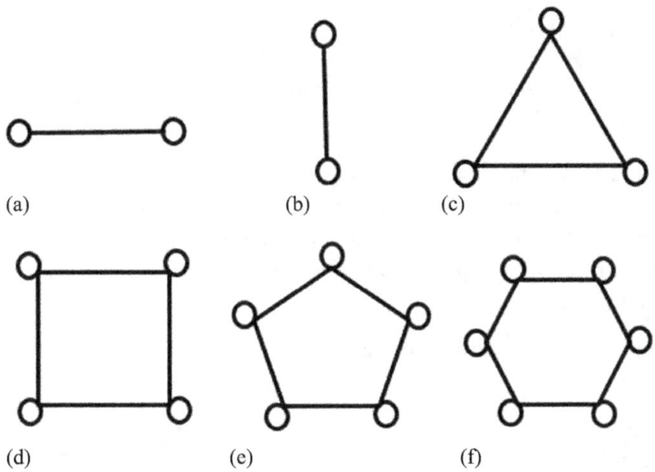

Figure 6.40 Different shapes of spaced damper used for bundled conductor (a) horizontal, (b) vertical, (c) triangle, (d) square, (e) pentagon, (f) hexagon

Figure 6.41 Teared jumper conductor in 33 kV line

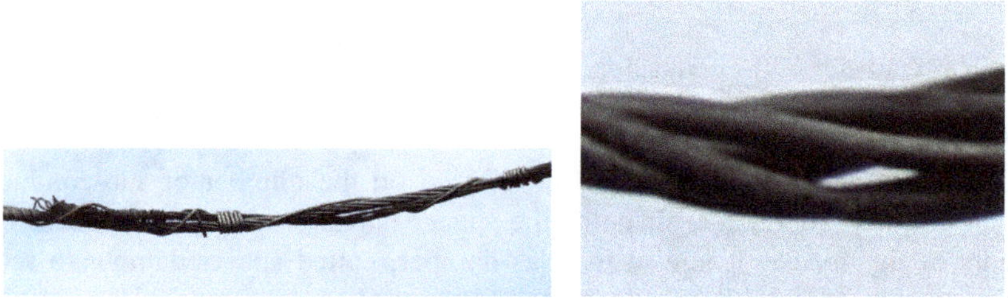

Figure 6.42 Birdcage in line conductor

6.18 Conductor damages

Major causes of conductor failures/damages (Figures 6.41 and 6.42) may be written as follows:

1. Failure due to mechanical motion of conductor
2. Failure due to unbalanced tension due to ice or wind or both
3. Failure due to fall of external bodies on the conductor
4. Electrical causes like short circuit, lightning
5. Fatigue and ageing effect

Different types of conductor damage noticed are as follows:

1. Tearing
2. Corrosion
3. Birdcage formation
4. Breaking of one or few strands

5. Break down conductor core
6. Joint failure
7. Failure in conductor fittings

Solutions of conductor damages are of two types:

1. Replacing
2. Repairing

6.19 Summary

Conductors used in overhead lines have been discussed in detail in this chapter. Materials used in overhead line conductor have been presented, and general classification of line conductors has been done. Electrical and thermomechanical design features of line conductor have been presented. Electrical properties, mechanical sag and thermal properties have been discussed. Selection criteria and fitting of conductor in the line have been presented. As conductors are exposed to open air, they undergo different types of vibration. Conductor vibration influences the tension of the line. All probable vibrations along with dampers have also been discussed.

6.20 Standards

Different standards for conductors have been presented in Tables 6.10–6.17.

Table 6.10 IEC Standard for conductor

IEC Standard (IEC)	Conductors
61089	Aluminium Aluminium alloy Aluminium clad steel ACSR ACSR/AC AACSR

Table 6.11 IEC Standard for wires

IEC Standard	Wires
60104	Heat treated 6000 series aluminium (AL3)
60121	Fully annealed 1350 aluminium (A0)
60889	Hard drawn 1350 H19 aluminium (AL1)
61089	Aluminium Aluminium alloy Aluminium clad steel Steel
62004	Thermal resistant zirconium aluminium (AT1) Super thermal resistant zirconium aluminium (AT1)

Table 6.12 European Standard for conductors

European Standard (EN)	Conductors
50182	Aluminium Aluminium alloy Aluminium clad steel ACSR ACSR/AC AACSR AACSR/AC
50189	Steel

Table 6.13 European Standard for wires

European Standard (EN)	Wires
50183	Aluminium alloy
50189	Steel
60889	Aluminium
61232	Aluminium clad steel

Table 6.14 German Standard for conductors

German Standard (DIN)	Conductors
48201 Part 1	Copper
48201 Part 2	Copper alloy
48201 Part 3	Steel
48201 Part 5	Aluminium
48201 Part 6	Aluminium alloy
48201 Part 8	Aluminium clad steel

Table 6.15 German Standard for wires

German Standard (DIN)	Wires
48200 Part 1	Copper
48200 Part 2	Copper alloy
48200 Part 3	Steel
48200 Part 5	Aluminium
48200 Part 6	Aluminium alloy
48200 Part 8	Aluminium clad steel

Table 6.16 ASTM Standard for conductor

ASTM Standards (ASTM)	Conductors
A363	Steel
B8	Aluminium alloy
B231	Aluminium
B232	ACSR
B341	ACSR/AC
B399	Aluminium alloy
B416	Aluminium clad steel
B711	AACSR

Table 6.17 ASTM Standard for wires

ASTM Standards (ASTM)	Wires
A475	Steel
B1	Copper
B105	Copper alloy
B230	Aluminium
B398	Aluminium alloy
B415	Aluminium clad steel

References

[1] O. K. Papailiou, Overhead Lines. CIGRE Green Books, Springer, Malters, Switzerland; 2017.

[2] Overhead Conductor Installation Guide Recommended Practices. 1st Edition, Installation guide by Electric Utility Engineering Section. General Cable Technologies Corporation, Kentucky, USA; 2014.

[3] C. G. Soares and C. Taralli, 'Insulated overhead lines'. in IEEE Transactions on Power Apparatus and Systems, vol. PAS-l01, no. 7, pp. 2273–2277, 1982.

Chapter 7
Earthing and earth wire

This chapter deals with earthing and earth wire used in overhead electric lines. Effect of electric current on human body has been mentioned. Resistance properties of soil have been described. Measurement techniques for earth resistance have been presented. Different earthing procedures and materials suitable for earthing have been presented. Limitation of ungrounded system and advantages of earthed system have been presented. Different types of system grounding have been presented. Fault current for different earthing systems have been described. Harmonic suppression system (HSS) and earthing transformer have been described. Common grounding practices have been mentioned. Different properties of earth wire and its selection guidelines have been presented. Design features of earth wires have been included. Practices for tower earthing have been described. Earthing for personal safety has also been described. Pipe earthing, counterpoise earthing, grounding in pole support, earthing of guard wire, etc. have been described.

7.1 Introduction

Earthing and earth wire are required for the safety of the power system network and the operating people under the line. Major goals of earthing and earth wire are as follows:

- Safety of the operating people
- Safety of the equipment
- To minimize damage due to various fault
- To help in detection of earth fault
- To limit in fault current in many cases
- To avoid economic loss

As overhead lines start from substation or generating station-connected substation, earthing of overhead lines relates with the earthing of substation from where it starts and to where it ends. This chapter will include

- Earthing of substation connected with overhead lines
- Earth-wire-connected-based earthing of overhead lines
- Other earthing systems used in overhead lines

7.2 Electric current on body

Cells and tissues decide electrical properties of a human body. Almost two-thirds part of human body consist of water and other liquids. If a power source is connected or touched with human body, flow of charge occurs. It depends on the resistivity of skin, muscles, etc. Dry skin can have resistivity up to 20–30 kΩ, but, resistivity of wet skin can decrease up to 0.5 kΩ, and raptured skin can have resistivity around 0.2–0.3 kΩ. For human risk, the following points are important:

- Current magnitude
- Voltage magnitude
- Skin resistivity
- The presence of moisture on skin
- Water content in the body
- Muscles' resistivity and charge withstanding capability
- Nerve's resistivity and charge withstanding capability strength
- Heart condition
- Platform of human body
- Sustainability of the electrical signals in the human body

For AC, 1 mA and for DC, 5 mA current may create electrical shock to the body. For 50–60 Hz AC, 50–70 mA and for DC, 250–500 mA current may be dangerous which may cause serious health injury or death.

7.3 Soil resistivity

Soil resistance is electrical property of soil. It refers to the ability by which soil opposes flow of current through soil. Soil resistivity depends on the following major factors:

- Chemical composition
- Moisture content in the soil
- Presence of salt in soil
- Temperature

The value of soil resistivity is not unique. It varies with different conditions. With the increase of temperature, soil resistivity decreases and vice versa. Soil resistivity decreases with the increase of moisture content in the soil. Soil resistivity also decreases with the increase of salt present in the soil.

Resistance is mathematically written as

$$R = \rho \frac{L}{A} \tag{7.1}$$

Soil resistance is measured in Ω cm or in Ω m.

If

$$L = 1\ \text{m},\ A = 1\ \text{m}^2,$$

then

$$R = \rho \tag{7.2}$$

Soil resistivity refers to the resistance of soil material between two sides of cubic volume of 1 m dimension at each side. Soil resistivity varies from 1.5 Ω m to 10,000 Ω m.

7.4 Electrode

Conductor embedded in soil for dissipating current in the earth is known as electrode. Electrodes are made of different shapes like rod and pipe, and they are embedded in soil in different fashions like horizontal and vertical. Electrodes are used to provide the path for the measurement of soil resistance and path for dissipating earth fault current of the system to earth.

7.5 Earthing mat or earthing grid

Metallic conductor in the form of grid ebbed in the soil is known as earthing mat or earthing grid. It is used to make uniform distribution of soil resistance and to reduce overall soil resistance in the area. Mat earthing covers wider area in soil and helps in easy flow of fault current.

7.6 Earthing conductor or earthing wire

The conductor which makes connection between earth electrode and the system to be earthed is known as earthing conductor or earthing wire. The dimension and rating of this conductor is designed on the basis of the fault current that can dissipate through the path during earth fault.

7.7 Materials used for earthing

Materials used in earthing vary with different areas of application. Materials used for earthing in soil, open air and in concrete has been presented in Table 7.1 [IEC-1024-1 (1990)]. As coating materials, lead and copper are used.

7.8 Touch potential

When a human body touches a surface of potential by finger, the potential difference measured between finger and step is known as touch potential. It is a measure of safety of the operating person with respect to the earthing of system.

Table 7.1 Materials used for earthing

Area	Materials commonly used
Soil	Copper (solid or stranded) Hot galvanized steel (solid) Lead (solid)
Open air	Copper (solid or stranded) Hot galvanized steel (solid or stranded)
Concrete	Copper (solid or stranded) Hot galvanized steel (solid) Lead (solid) Aluminium (solid or stranded)

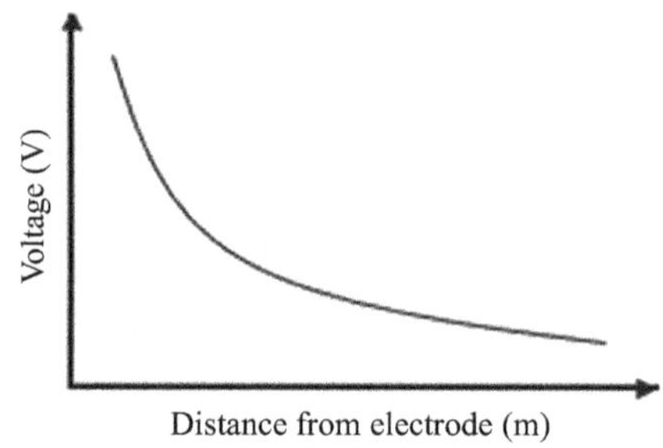

Figure 7.1 Voltage gradient

7.9 Step potential

When a human body stands on earth surface of a substation, the potential difference measured between two foots separated by a step is known as step potential. It is also a measure of safety of the operating person with respect to the earthing of system.

7.10 Voltage gradient

Voltage decreases with the increase of distance from earth electrodes. Voltage gradient nature normally found in soil has been shown in Figure 7.1.

7.11 Soil resistance and its measurement

Resistance offered by soil between specific distances of electrodes embedded in specific depth is known as soil resistance. It is measured in conventional way by measuring the ratio of voltage and current. Potential is created across electrode and let current flows through the soil. Let, V =voltage and I=current.

Then earth resistance will be

$$R_{th} = \frac{V}{I}$$

7.12 Soil resistance measurement

Soil resistance is conventionally measured by three ways as follows:

- Wenner four pole equal method
- Schlumberger array method
- Driven rod method

Circuit diagrams showing different arrangements of placing electrodes for the previous different methods of measurement of soil resistivity have been shown in Figure 7.2(a)–(c), respectively. Soil resistivity measured by different measurement methods has been presented in Table 7.2.

Schlumberger array method gives better result compared to Wenner four pole equal method and driven rod method. Also, Schlumberger array method needs less labours' effort than other two methods.

7.13 Earth resistance of electrode and its measurement

Electrode, being made of conducting material, while embedded in soil, faces a resistance between the electrode itself and earth (sometimes referred to as true earth). Electrode is characterized by the resistivity of the material used in it and earth is characterized by the soil resistivity. The resistance between an electrode and true earth is known as earth resistance. Commonly used methods for the measurement of earth resistance of electrodes are as follows:

- Single vertical-rod-driven method
- Parallel-rod-driven method
- Horizontal trench-electrode-driven method

Earth resistance of electrodes as measured by these three methods has been presented in Table 7.3.

Earth resistance of an electrode may be measured in the following ways in addition to the previous methods:

- Fall of potential method
- 62% method
- Slope method
- Star delta method
- Four potential method

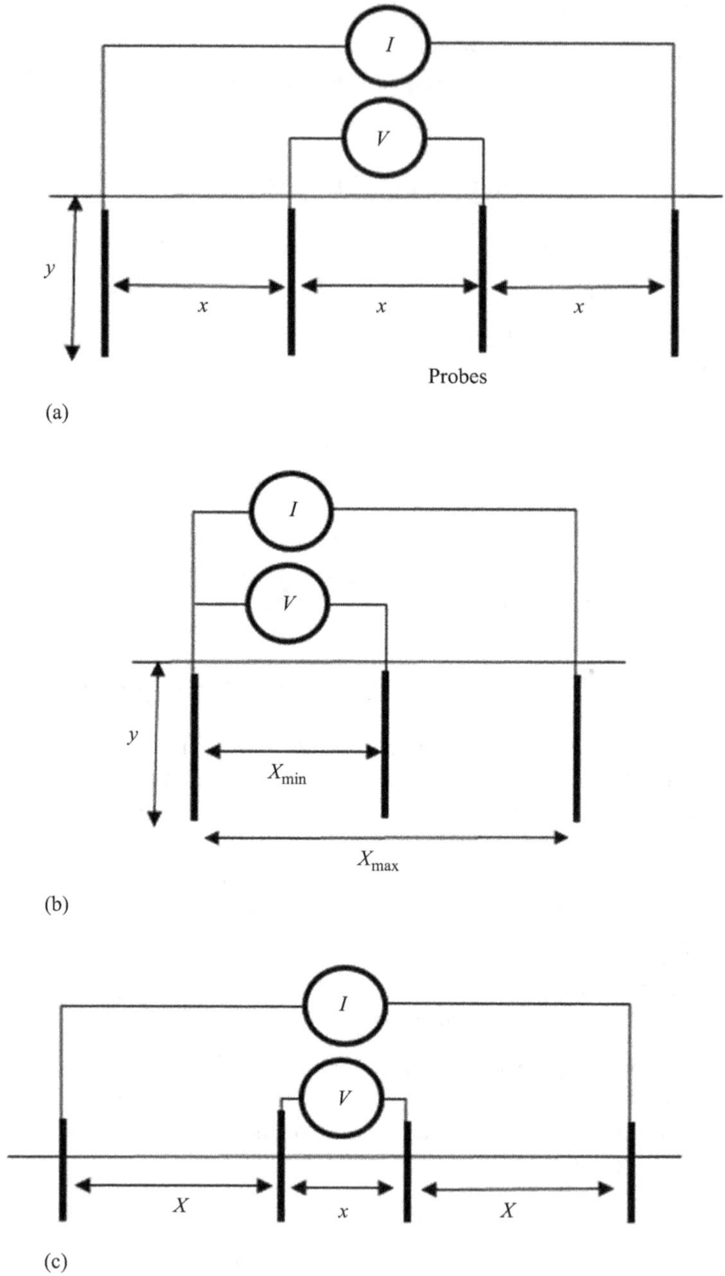

Figure 7.2 Soil resistivity measurement by (a) Wenner four pole equal method, (b) Schlumberger array method and (c) driven rod method

Table 7.2 Soil resistivity measured by different measurement methods

Method	Resistivity	Number of electrodes required	Parameters
Wenner four pole equal method	$\rho_s = \dfrac{4\pi xR}{1+\left(\dfrac{2x}{\sqrt{x^2+4y^2}} - \dfrac{x}{\sqrt{x^2+y^2}}cc\right)} \cong 2\pi xR$	4	R = measured resistance, x = distance between two consecutive electrodes, y = depth of electrode in soil
Schlumberger array method	$\rho_s = \frac{\pi(X^2-x^2)}{4x}\frac{\Delta V}{I}$	4	X = distance of outer probe from inner probes, x = distance between inner probes, ΔV = voltage difference, I = current
Driven rod method	$\rho_s = \frac{2\pi l_{Rod}R}{\ln\frac{8l_{Rod}}{d_{Rod}}}$	3	R = measured resistance, l_{Rod} = length of rod, d_{Rod} = diameter of rod

Table 7.3 Earth resistance of electrodes as measured by these three methods

Method	Resistivity	Number of electrodes	Parameters
Single vertical rod-driven method	$R_{EL} = \frac{\rho_s \ln\left(\frac{8L_{Rod}}{d_{Rod}} - 1\right)}{4\pi L_{Rod}}$	1	ρ_s = soil resistivity, d_{Rod} = diameter of electrode, L_{Rod} = length of electrode buried in soil
Parallel-rod-driven method	$R_{EL} = \frac{\rho_s \ln\left(\frac{2L_{Rod}}{r_{Rod}} - 1\right)}{\pi L_{Rod}}$	>1	ρ_s = soil resistivity, r_{Rod} = radius of electrode above soil, L_{Rod} = length of electrode buried in soil
Horizontal trench-electrode-driven method	$R_{EL} = \frac{\rho_s \ln\left(\frac{4L_{Rod}}{\sqrt{d_{trench}h_{trench}}} - 1\right)}{\pi L_{Rod}}$	1	ρ_s = soil resistivity, d_{trench} = diameter of electrode, L_{trench} = length of electrode buried in soil, h_{trench} = length of electrode buried in soil

Table 7.4 Different values of K_S for different numbers of electrodes in star connection

Number of electrodes in star connection	K_S
2	1.03
3	1.06
4	1.12
5	1.42
6	1.65

7.14 Radial or star connection of electrodes

Often more than one number of electrodes are connected in radial or in star shape. Star can be made by multiple numbers of electrodes like 3, 4, 5, 6, etc. This effectively increases the resistance of electrode. Let, R=resistance of an electrode.

Then effective resistance of electrodes in radial or in star connection will be

$$R_{Eff} = K_S R \tag{7.3}$$

where K_S is the multiplying factor for star connection and is greater than 1. Different values of K_S for different number of electrodes in star connection have been presented in Table 7.4.

7.15 Limitation of isolated neutral or ungrounded system

Main limitations of an ungrounded system are as follows:

1. Neutral shifting occurs during unbalanced loading of fault in ungrounded system.
2. During fault, fault current flows through healthy phases in ungrounded system.
3. During fault, voltage levels of healthy phases increase in ungrounded system.
4. Often first short circuit fault is difficult to detect in ungrounded system. Fault detection gets complicated.
5. It is difficult to limit fault current in ungrounded system.

7.16 Neutral grounded system

The power system where the neutral point is forcefully grounded and kept at near zero potential is known as grounded system.

Main advantages of an ungrounded system are as follows:

1. Neutral shifting does not occur during unbalanced loading of fault in grounded system.

2. During fault, fault current does not flow through healthy phases in grounded system.
3. During fault, voltage levels of healthy phases do not change in grounded system.
4. Fault detection becomes simple in grounded system.
5. It is easy to limit fault current in grounded system.
6. Proper grounding reduces voltage stress during steady state condition.
7. Sensitivity of protective devices increase in grounded system.
8. Proper grounding helps in transients and high voltage protection.
9. Grounding increases system reliability.
10. Safety level to the operating persons increases.

7.17 Different grounding methods

There are different types of grounding methods available in power system. The choice of grounding method depends on the system requirement, allowable loss and cost to be spent for grounding purpose. Grounding methods are mainly classified into three categories as follows:

1. Solid grounding
2. Resistance grounding
3. Reactance grounding

Other than the previous methods, grounding can be done by arc suppression coil, voltage transformer and earthing transformer. Different grounding methods with merits and demerits are discussed in the following sections.

7.17.1 Solid grounding

In this method, the neutral point is grounded directly to the ground. The conductor bar used for this purpose has very low, effectively negligible resistance.

A typical solid grounded system is shown in Figure 7.3.

Merits: The main advantages of solid grounding are as follows:

1. The method is simple.
2. Voltage drop in the grounding wire is zero.
3. Cost is very small.

Demerits: The main disadvantages are as follows:

1. Fault current is not limited by solid grounding.
2. Solid grounded system requires higher setting of relay and higher ratings of circuit breaker.

7.17.2 Resistance grounding

In this method, the neutral point is grounded through a resistance to the ground. Thus, the grounding path has considerably high resistance. By proper choice of the resistance, fault current can be limited by resistance grounding. Mathematically, if

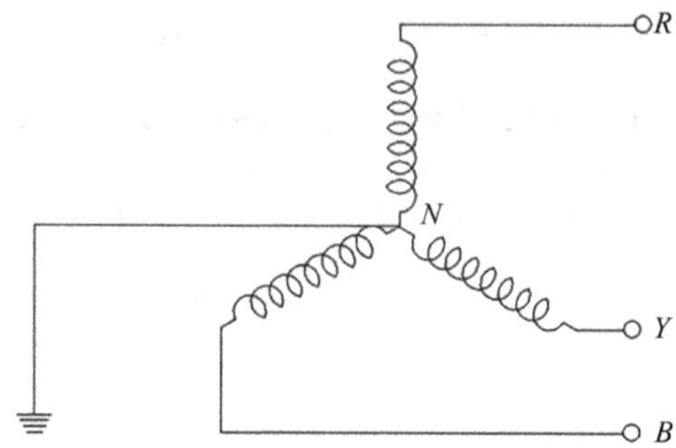

Figure 7.3 Schematic diagram of solid grounded system

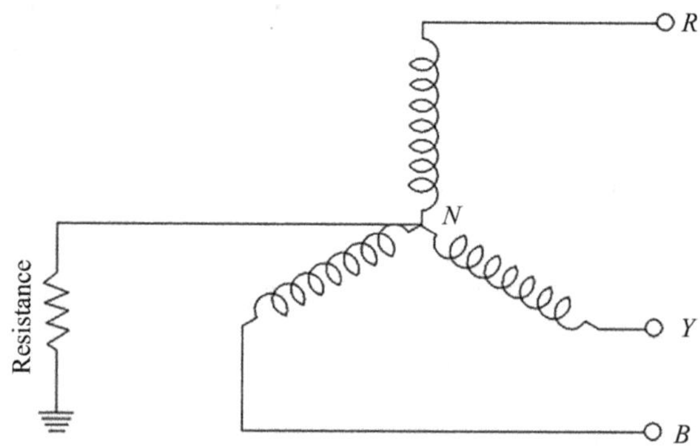

Figure 7.4 Schematic diagram of resistance-grounded system

V be the voltage level and X be the total reactance in the fault path, then the fault current would be

$$\text{Fault current } (I_F) = \frac{V}{X}$$

The introduction of resistance (R) by the resistance grounding system reduces the fault current and it is given as

$$\text{Fault current } (I_F) = \frac{V}{R+X}$$

A typical resistance-grounded system is shown in Figure 7.4.

Merits: The main advantages of resistance grounding are as follows:

1. The method is less complicated with respect to reactance grounding.
2. Fault current is limited by grounding resistance.
3. Cost is comparatively small with respect to reactance grounding.

Demerits: The main disadvantages are as follows:

1. The method is more complicated with respect to solid grounding.
2. Grounding resistance increases the copper loss when fault current flows through it.
3. Cost is higher than solid grounding.

7.17.3 Reactance grounding

In this method, the neutral point is grounded through a reactance to the ground, i.e. the grounding path is highly reactive. By proper choice of the reactance, fault current can be limited by reactance grounding. Mathematically, if V is the voltage level and X is the total reactance in the fault path, then the fault current would be

$$\text{Fault current } (I_F) = \frac{V}{X}$$

The introduction of reactance (X_G) by the reactance grounding system reduces the fault current and it is given as

$$\text{Fault current } (I_F) = \frac{V}{X_G + X}$$

A typical reactance-grounded system is shown in Figure 7.5.

Merits: The main advantages of reactance grounding are as follows:

1. Fault current is limited by grounding resistance.
2. As the grounding path has very low resistance, copper loss due to fault current passing through the ground path is almost negligible.

Demerits: The main disadvantages are as follows:

1. The method is more complicated with respect to other methods.
2. Cost is higher than solid grounding.

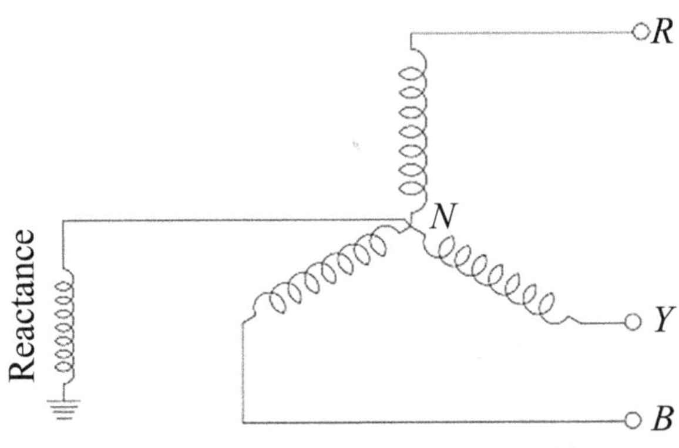

Figure 7.5 Schematic diagram of reactance-grounded system

7.17.4 Grounding by arc suppression coil

In this method, neutral of a three-phase electric power system is grounded by a coil having variable reactance as shown in Figure 7.6. The coil is known as Peterson's coil. Reactance of the coil can be varied by changing the number of turns with the help of tapping. Reactor limits fault current and its value is adjusted by tap changer. It is safe and suitable for the system.

7.17.5 Grounding by voltage transformer

Grounding by voltage transformer has been shown in Figure 7.7. A voltage transformer is connected in the ground path. Secondary of the transformer is connected with protective relay for fault detection.

7.17.6 Comparison

Comparison between solid, resistance and reactance groundings has been presented in Table 7.5.

7.18 Resonant grounding or Peterson coil grounding

Line conductor forms capacitor with earth. In ungrounded system or isolated neutral system, earth fault results in capacitive current. In reactance grounding, reactance is inserted in between neutral and earth. This provided parallel inductive path to the capacitive path of fault current. If the reactance is adjusted in such a way that it opposes completely capacitive current and nullifies each other, then it is called resonant grounding. The coil which provided inductive reactance in the path of fault current is known as Peterson coil, and sometimes it is referred to as Peterson coil grounding. If inductive current exceeds capacitive current, it is called over resonant and if inductive current is less in magnitude than capacitive current, then it is called under resonant. Coil and capacitor may have resistive loss which contributes as resultant resistive current. If they are assumed lossless, then the current at resonant grounding will be zero.

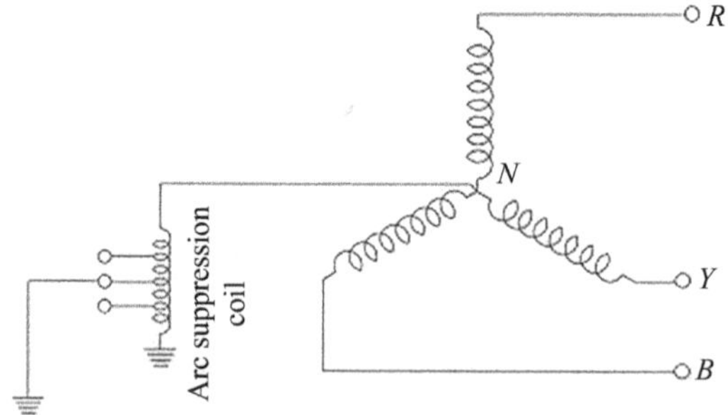

Figure 7.6 Schematic diagram of neutral grounding by arc suppression coil

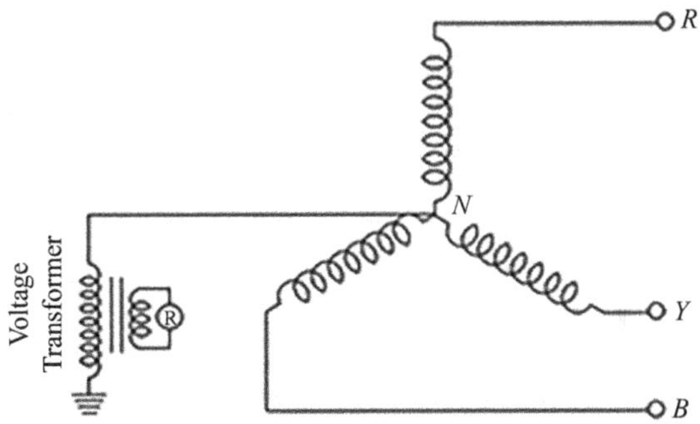

Figure 7.7 Schematic diagram of neutral grounding by voltage transformer

Table 7.5 Comparison between solid, resistance and reactance groundings

No.	Solid grounding	Resistance grounding	Reactance grounding
1	Most simple	Complicated	Complicated
2	Fault current cannot be limited	Fault current can be limited	Fault current can be limited
3	Copper loss is small	It has high copper loss during the flow of fault current	Copper loss is small
4	Cost is small	Cost is higher than resistance grounding and less than reactance grounding	Cost is highest

7.19 Fault current at different earthing system

Current flowing through the fault path depends on the type of earthing. Estimation of fault current is essential for accurate and reliable operation of relay, switchgear and other protective devices. Fault currents in unearthed and earthed system have been discussed in the following subsections.

7.19.1 Fault current at isolated neutral or unearthed system

Conductors of overhead line form capacitor with earth. If earth fault occurs in isolated neutral system, then ground fault current is flown through capacitor formed between healthy phase and earth. Let C be the capacitance of the capacitor formed in between each phase conductor and earth. Let a ground fault has occurred at phase B as shown in Figure 7.8. Then fault current through phase B will flow through capacitor formed between two healthy phases R and Y. Therefore, fault current may be mathematically expressed as

$$I_F = I_C = I_{RC} + I_{YC} \tag{7.4}$$

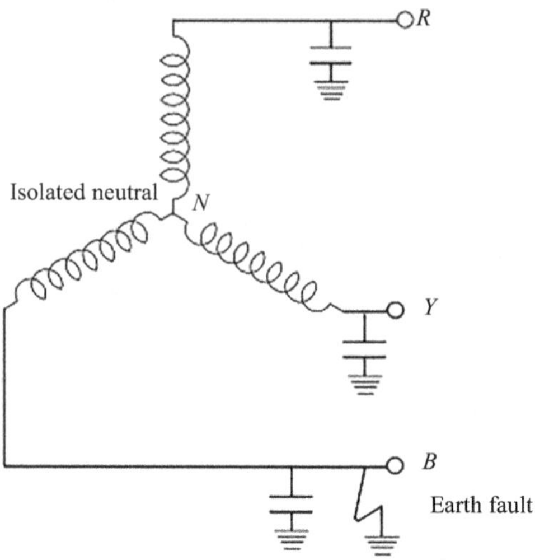

Figure 7.8 *Single line to ground 'earth fault' in isolated neutral three-phase system*

7.19.2 Fault current at resistance earthed system

Resistance-grounded system has been shown in Figure 7.9. As conductors of overhead line form capacitor with earth, if earth fault occurs in resistance earthed system, then ground fault current is flown through capacitor formed between healthy phase and earth, and in addition to that resistance in the earthling system will provide parallel resistive path to the capacitive current. Let C is the capacitance of the capacitor formed in between each phase conductor and earth, R_G is the resistance in the earthing system and R_F is the resistance in the fault path. Consider a ground fault has occurred at phase B. Then fault current through phase B will flow through ground resistance as well as through capacitor formed by two healthy phases R and Y. Therefore, fault current may be mathematically derived as follows:

$$I_F = I_C + I_G = (I_{RC} + I_{YC}) + I_R \tag{7.5}$$

Let, $X_C =$ capacitive reactance formed by each phase with the earth and V is the voltage between phase and neutral.

$$I_{RC} = \frac{\sqrt{3}V}{X_C} \tag{7.6}$$

$$I_{YC} = \frac{\sqrt{3}V}{X_C} \tag{7.7}$$

$$I_F = I_C + I_G = I_C + I_R = \frac{3V}{X_C} + I_R = j3\omega CV + \frac{V}{R_G} \tag{7.8}$$

Based on resistance value, it is divided into two categories: low-value and high-value resistance grounding. However, there is no clear demarcation between these two categories in terms of magnitude of resistance. Low-value grounding

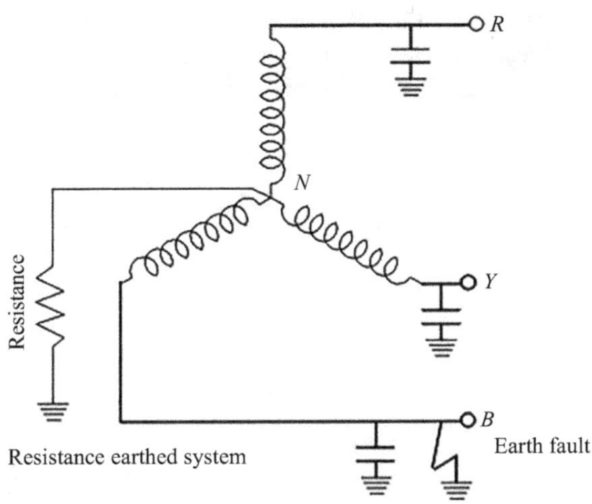

Figure 7.9 Single line to ground 'earth fault' in resistance earthed three-phase system

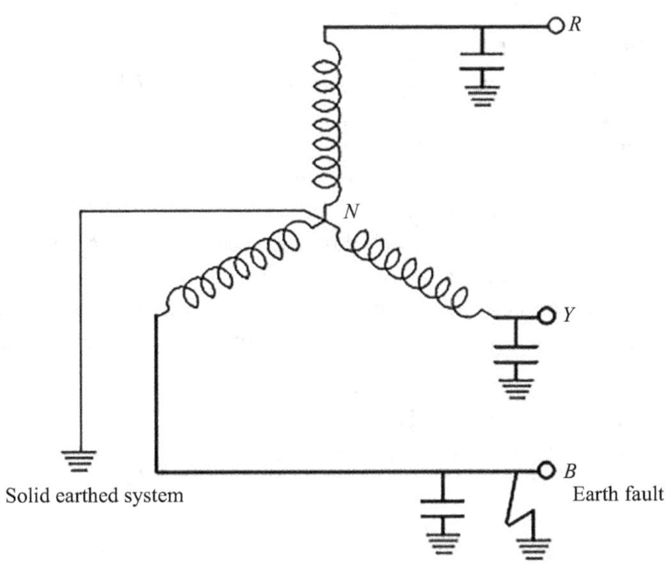

Figure 7.10 Single line to ground 'earth fault' in solid earthed three-phase system

results in high earth fault current, whereas high-value resistance grounding results in comparatively low earth fault current.

7.19.3 Fault current at solid earthed system

Solid grounded system has been shown in Figure 7.10. In solid grounding, R_G, the resistance in the earthing system, is very small and R_F is the resistance in the fault path.

Let a ground fault has occurred at phase B. Then fault current through phase B will flow through ground resistance as well as through capacitances of two healthy phases R and Y. Therefore, fault current may be mathematically expressed as follows:

$$I_F = I_C + I_G \tag{7.9}$$

where,

$$I_C = I_{RC} + I_{YC} \tag{7.10}$$

$$I_G = I_R = \frac{V}{R_G} \tag{7.11}$$

For solid earthed system,

$$R_G \ll X_C$$

Therefore,

$$I_R \gg I_C$$

$$I_F \approx I_R \tag{7.12}$$

7.19.4 Fault current at reactance earthed system

Reactance-grounded system has been shown in Figure 7.11. As conductors of overhead line forms capacitor with earth, if earth fault occurs in resistance earthed system, then ground fault current is flown through the capacitor formed between healthy phase and earth and in addition to that reactance in the earthling system will provide parallel inductive path to the capacitive current.

Let C be the capacitance formed in between each phase conductor and earth, X_{LG} is the reactance in the earthing system and R_F is the resistance in the fault path. Consider a ground fault has occurred at phase B. Then fault current through phase B will flow through ground inductance as well as through capacitances of two healthy phases R and Y.

a) Fault current without resistive loss

Fault current without resistive loss may be mathematically expressed as follows:

$$I_F = I_C + I_G$$

where,

$$I_C = I_{RC} + I_{YC} \tag{7.13}$$

For reactance earthed system,

$$I_G = \frac{V}{X_G}$$

$$X_G = j\omega L_G = X_L \text{ (say)} \tag{7.14}$$

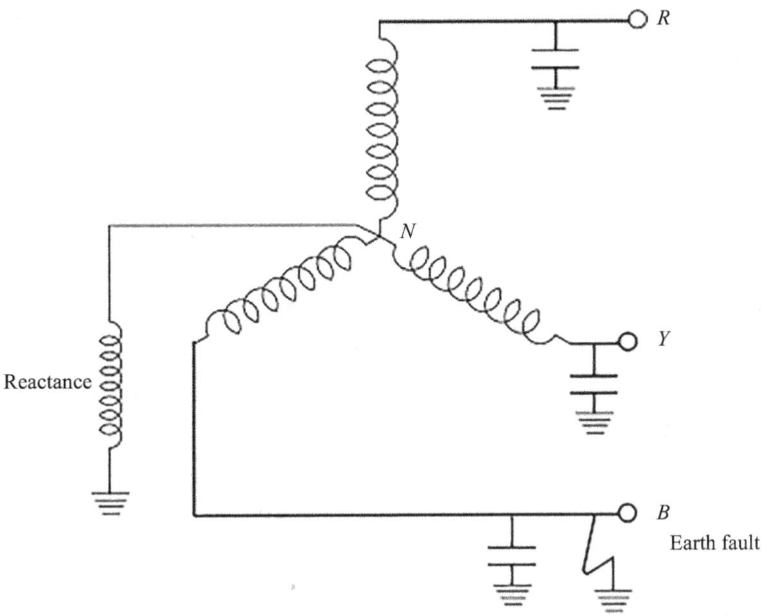

Figure 7.11 Single line to ground 'earth fault' in reactance earthed three-phase system

$$I_G = \frac{V}{X_L} = \frac{V}{j\omega L_G} = I_L \tag{7.15}$$

$$I_F = I_C + I_L \tag{7.16}$$

b) Fault current with resistive loss

Considering resistive loss in capacitive and inductive path, fault current may be mathematically expressed as follows:

$$I_F = I_C + I_G + I_{CR} + I_{GR} \tag{7.17}$$

where,

$$I_{CR} = \frac{V}{R_C} \tag{7.18}$$

$$I_{GR} = \frac{V}{R_L} \tag{7.19}$$

$$\begin{aligned} I_F &= I_C + I_L + I_{CR} + I_{GR} \\ &= \frac{3V}{X_C} + \frac{V}{X_L} + \frac{V}{R_C} + \frac{V}{R_L} \\ &= j3\omega CV + \frac{V}{j\omega L_G} + \frac{V}{R_C} + \frac{V}{R_L} \end{aligned} \tag{7.20}$$

7.19.5 Resonant earthed system

At resonance,

$$I_C + I_L = 0 \tag{7.21}$$

or,

$$I_C = -I_L \tag{7.22}$$

Condition for resonance is

$$\frac{3V}{X_C} + \frac{V}{X_L} = 0 \tag{7.23}$$

or,

$$\frac{3V}{X_C} = -\frac{V}{X_L} \tag{7.24}$$

Therefore,

$$|X_L| = \frac{X_C}{3} \tag{7.25}$$

a) **Fault current considering resistive loss**
Considering resistive loss in earth coil and capacitive path as shown in equivalent circuit for single line to ground fault, fault current may be mathematically expressed as follows:

At resonance,

$$I_F = I_C + I_L + I_{CR} + I_{GR} = 0 + \frac{V}{R_C} + \frac{V}{R_L} = \frac{V}{R_C} + \frac{V}{R_L} \tag{7.26}$$

If R_F be the resistance at fault path, then fault current at resonant will be

$$I_F = \frac{V}{R_F + \left(\frac{R_C R_L}{R_C + R_L}\right)} \tag{7.27}$$

b) **Fault current neglecting resistive loss**
Neglecting resistive loss in earth coil and capacitive path for single line to ground fault, fault current may be mathematically expressed as follows:

$$\begin{aligned} R_L &= \infty \\ R_C &= \infty \\ I_{CR} &= 0 \\ I_{GR} &= 0 \end{aligned}$$

At resonance,

$$I_C + I_L = 0$$

Therefore,

$$I_F = I_C + I_L + I_{CR} + I_{GR} = 0 \tag{7.28}$$

If R_F be the resistance at fault path, then fault current at resonant will be

$$I_F = \frac{V}{R_F} \tag{7.29}$$

7.20 Harmonic suppression system

Electric power system faces different types of harmonics mainly due to non-linear devices, non-linear loads and use of semiconductor-made switches. Odd harmonics are very common. Transformer inrush causes even harmonics, whereas semiconductor-based switching units generate lots of inter- and subharmonics. Harmonics contribute loss as well as malfunction of controlling and protective devices. Third harmonics (and triplen) are very common that increase current flow through neutral or ground line causing the generation of voltage drops in the ground line. This gives rise of zero sequence components. These harmonics, if not considered, can cause mal operation of the relay unit. Harmonic suppression system (HSS) eliminates harmonics (mainly third harmonics) from the ground paths. Needs and application of harmonics suppression systems are found widely in distribution system.

7.21 Earthing transformer

Earthing transformer is special type of transformer used to provide neutral point for earthing which does not have normal neutral point of star connection. Part of power system network having no neutral point is grounded with the help of earthing transformer as shown in Figure 7.12. This transformer does not carry electrical loads and provides neutral point by its secondary common points of secondary windings that is grounded by different methods. Its volt-ampere (VA) rating is made very small to meet the loss components only. As purpose of earthing transformer is not to step up or step down the voltage, its turns ratio is normally made as 1:1.

7.22 Grounding practice

Overhead electric lines must be designed, installed and maintained abiding by the grounding practices as prescribed by the regulatory body of the respective country or region. Safety is given utmost important followed by the protective measure for the lines, towers and other connected members of overhead lines. Guidelines given by IEEE and other societies are followed. If there is no advice from regulatory body, then it is advisable to follow guidelines by IEEE or other directly for safety and secure operation of the system. IEEE standard 1048-1990 for guide for protective grounding of power lines has been revised in 2003 and 2016 in this regard.

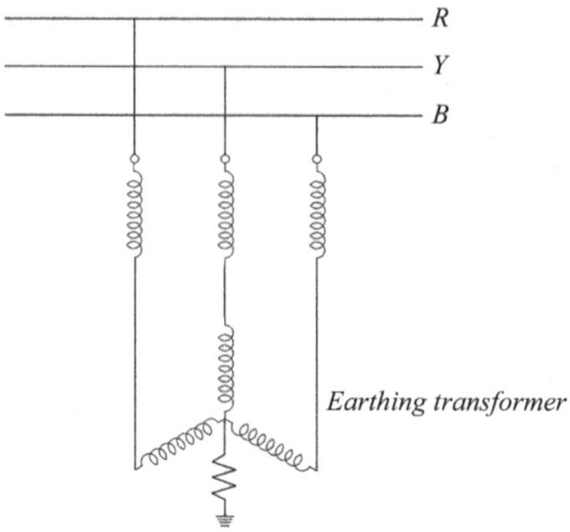

Figure 7.12 Earthing transformer

Individual attention must be given for grounding of the following parts in overhead line:

- Grounding of sky wire
- Grounding of tower body and legs
- Grounding neutral points
- Grounding of support ends of line insulators
- Grounding of guys
- Grounding of guard wires

7.23 Earthing for personal safety

Personal safety for the persons operating or handling electrical systems or sub-system is of utmost importance. For equipment, outer body of the equipment that may be in touch with the person must be brought into zero potential by earthing the body. If earth fault occurs, fault current will flow through the ground path keeping the voltage level at ground level. If anybody touches the equipment, no current will flow through him ensuring his safety. In overhead lines, all metallic parts of the path except line conductors and its junction points must be grounded. Separate continuous earth wire is provided. In lattice tower, the structure is ensured at ground potential. One, two or four earthing is made. All metallic parts of support ends of line insulators, cross arm, guy, guard wire and the sky wire are ensured at ground potential.

7.24 Earth wire

Earth conductor is installed for the overhead line, connected power system network and the safety of the people under the line. It is directly connected to the tower and grounded. Earth wire is placed above the phase conductor at the top of the tower. For this reason, it is also called sky wire. The number of earth conductor (1 or 2 or higher) depends on the voltage level of the overhead line and height of the tower. Continuous operating current is negligible with respect to phase conductor but increases during faults. Main purposes of the earth wire or sky wire are:

- to provide protection against lightning surges.
- to pass the surge current as early as possible.
- to resist the induction of surges due to lightning in the phase conductors placed at the cross arms of the tower below the top.

7.25 Design features of earth wire

Earth wire provides path for fault current to the ground. Therefore, design of earth wire depends on the maximum current expected to pass through this wire is very important. Accordingly, temperature status is considered. Two following phenomena occur during fault:

- Heat generation due to fault dissipation to ground.
- Heat dissipation through surface area.

Now, in the overhead lines above 110 kV voltage levels, auto-reclosing facility is provided. If fault clears by that time, after auto-reclosing, only heat dissipation takes place. But if fault remains ever after auto-reclosing, heat generation will continue; heat dissipation may be neglected as long as high fault current persists.

Materials used in earth wire must withstand heat maintaining the tensile strength profile within limit of stability.

Tensile strength, slack and sag characteristics of earth wire should be such that the resultant sag of earth wire should be less than sag of phase conductors.

When lightning-stroke occurs, overhead ground wire (OHGW) provides shunt path for the current to ground. The current passes through the impedance offered by the structure and footing resistance. In this way, OHGW reduces voltage stress across line insulator used for phase conductors and the probability of flashover. Ground wire removes the strokes in harmless way.

In low-voltage distribution lines, neutral is grounded and placed above the phase conductor that provides protection against lightning.

Shielding angle with respect to overhead ground line refers to the angle produced with outer most phase conductor and vertical line passing through it. Shielding angle may be of following two categories:

- positive angle as shown in Figure 7.13(a)
- negative angle as shows in Figure 7.13(b)

Earlier (before 1951) shield angle was used to set as 30° for line up to 250 kV. With the advancement of technology, a lot of research works have been carried out to select proper best fit shield angle. Various models have been introduced. Some of them are as follows:

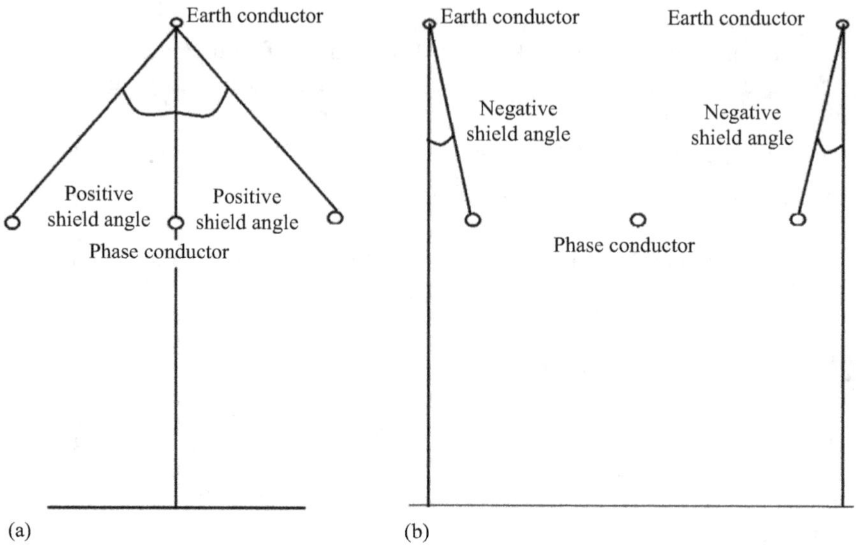

Figure 7.13 Shield angle: (a) positive and (b) negative

- Electromagnetic models
- Eriksson's model
- Generic model
- Statistical model

Striking distance: It refers to the distance from ground level where lightning stroke takes place.

Attraction radius: Distance from the ground wire up to which it can shield properly.

Selection of shield angle as described in most of the model depends on the following factors:

- Height of phase conductor
- Height of earth wire
- Radius of attraction

Shield angle obtained by different models varies. However, much lesser (than 30°) is used as shield angle.

7.26 Earth wire selection

Selection of the earth wire for overhead transmission lines is based on the matching of the following properties of conductor:

- Current carrying capability or ampacity or thermal rating
- Maximum operating conductor current

- Weight and allowable stress strength limit
- Height of the tower and voltage level of the overhead lines

7.27 Optical ground wire fibre reinforced

Conductors with optical fibre cable are special type of conductors where optical fibre cable is placed in core as shown in Figure 6.10 of Chapter 6. This cable is made of communication of high-frequency signals. Application of conductors with optical fibre cable is very limited on overhead lines but has future potential particularly in earth wire where load at normal condition is very low and can benefit communication for fault information as well as for other purpose. In protection scheme, this can be used as additional conductor to carry fault information.

7.28 Earthing of tower

Lattice towers are made of metallic body with concrete foundation. Legs of the tower must be grounded. Commonly used grounding methods are as follows:

- Pipe earthing
- Counterpoise earthing

7.28.1 Pipe earthing

Pipe earthing can be provided to all legs. In practice, at least one leg must be grounded. Any one leg can be chosen for this purpose. However, the leg to be grounded is fixed by the designer and marked by some specific symbol. For river crossing tower at least two (diagonally located legs) must be grounded. Pipe buried below the ground level is connected with the leg through galvanized steel tape. Surrounding of the pipe is filled by charcoal and salt to reduce soil resistance.

7.28.2 Counterpoise earthing

In this case, galvanized wire is connected with tower leg by lug, nut and bolt arrangement. The wire should be properly tightened with lug by nuts and bolts. In good practice, all the four legs are connected with wire and grounded. Wire was also buried below the ground level. In rock soil, care should be taken to avoid damage of the wire.

7.29 Grounding in pole support

Pole used in overhead lines is made of wood, steel or concrete. Normally, in distribution lines sky wire is not provided. However, all non-current carrying parts of the support should be grounded. In practice, grounding must be provided at the interval of at least four poles and at both ends of the line at substation. Pipe grounding with the help of galvanized iron wire connected with pole through lug, nuts and bolts arrangement is commonly used. Charcoal and salt is filled in the soil surrounding the pipe buried below the ground level to reduce earth resistance of the electrode.

In the case of non-metallic pole like concrete pole, separate earth wire is connected with pole and goes to the top of the pole; wire being tightened with pole by clamp arrangement. Sometime, such separate earth wire is provided for steel poles also. Guard lines and insulator base end (opposite end of line conductor) must be grounded.

7.30 Earthing of guard wire insulators' support end

End terminal of insulators installed on line support are connected with phase conductors. Starting terminals are based on support in various form with or without cross arm. This end must be grounded.

Guard wires are provided for safety in the line between two supports. When the line crosses road, rails, populated areas, etc., guard wires must be grounded. Guard wires are connected with the cross arm and is connected with earth wire which makes the connection with earthing system of the support. Support made of lattice tower provides ground path by the structure itself. But for the poles (like concrete pole) where structure does not provide path to earth, separate earth wire is connected for grounding. These earth wires make the connection from support end of insulators to the earth.

7.31 Neutral grounding in LV distribution line

As low-voltage distribution system distributes electric power to three-phase as well as single-phase users, it must carry neutral conductor along with phase conductor that comes from the substation. Neutral line at substation must be grounded. In addition, each support must have its own earth wire located at pole site.

7.32 Research advancement

With the advancement of material technology, earthing system in overhead lines has been advanced. Lots of research works are going on to provide better earthing to the overhead lines for safe and secure operation. Different design considerations and earthing methods have been advanced [1–3] and new guidelines are coming. Some useful definitions have been provided in [4] which research can consider for farther study. Changes in the grounding technologies have been directed in [5–8]. Shielding has become important for safety measures. Better shield technologies have been discussed in [9–11]. In addition to that, relevant standards have been introduced for research and practice considerations.

7.33 Summary

For safe and secure operation of overhead lines, earthing is necessary. In this chapter, methods of earthing and their application have been presented. Earth wire or sky wire has been discussed and selection criteria have been mentioned. Limitation of ungrounded system and advantages of grounded system have also been provided. Different grounding techniques have been presented mentioning their applications.

7.34 Standards and guidelines

Some useful standards and guidelines have been presented in Table 7.6 for further reading.

Table 7.6 Some useful standards and guidelines

Standard	Purpose
IEEE Std 81-2012	IEEE guide for measuring earth resistivity, ground impedance, and earth surface potentials of a grounding system
IEEE Std 524a-1993	IEEE guide to grounding during the installation of overhead transmission line conductors
IEEE Std 524-2003	IEEE guide to the installation of overhead transmission line conductors
IEEE Std 691-2001	IEEE guide for transmission structure foundation design and testing
IEEE Std 1025-1993	IEEE guide to the assembly and erection of concrete pole structures
IEEE Std 1048-1990	IEEE guide for protective grounding of power lines
IEEE Std 1070-2006	IEEE guide for the design and testing of transmission modular restoration structure components
IEEE Std. 1243-1997	IEEE guide for improving the lightning performance of transmission lines
IEC 60364-1 Fifth edition 2005-11	Low-voltage electrical installations – Part 1: Fundamental principles, assessment of general characteristics, definitions
IEC 60826:2017	Design criteria of overhead transmission lines
Indian Standard-5613 (BIS), 1996	Code of practice for design, installation and maintenance of overhead power lines
Indian Standard-2121 (BIS), 1991	Conductor and earth wire accessories for overhead lines
National Electric Safety Code (NESC), 1997 Edition, in C2-1997	National Electric Safety Code (NESC)

References

[1] K. O. Papailiou, Overhead Lines. CIGRE Green Books, Springer, Malters, Switzerland; 2017.

[2] Overhead Conductor Installation Guide Recommended Practices. 1st Edition, Installation guide by Electric Utility Engineering Section. General Cable Technologies Corporation, Kentucky, USA; 2014.

[3] F. Kiessling, P. Nefzger, J. F. Nolasco, and U. Kaintzyk, Overhead Power Lines, Springer, Berlin, 2003; ISBN: 978-3-642-05556-0.

[4] G. Cafaro, P. Montegiglio, F. Torelli *et al.*, 'The global grounding system: definitions and guidelines'. 2015 IEEE 15th International Conference on Environment and Electrical Engineering (EEEIC), Rome, 2015, pp. 537–541, doi: 10.1109/EEEIC.2015.7165219.

[5] H. Makino and H. Itakura, 'Study of overhead ground-wire shunting rates during grounding failure, for 500 kV substation grounding system design'. IEEE Power Engineering Society. 1999 Winter Meeting (Cat. No. 99CH36233), New York, NY, USA, 1999, pp. 994–999, vol. 2, doi: 10.1109/PESW.1999.747334.

[6] IEEE Std C62.92.2-2017 (Revision of IEEE Std C62.92.2-1989): IEEE Guide for the Application of Neutral Grounding in Electrical Utility Systems, Part II—Synchronous Generator Systems; 19 May 2017, pp. 1–38, doi: 10.1109/IEEESTD.2017.7932206.

[7] H. Pettersson and I. Pohjonen, 'Power calculations and grounding in EDP/UPS power networks'. 10th International Telecommunications Energy Conference, San Diego, CA, USA, 1988, pp. 494–499, doi: 10.1109/INTLEC.1988.22397.

[8] I. Darney, 'Grounding, floating and screening'. IEE Seminar on Shielding and Grounding (Ref. No. 2000/016), London, UK, 2000, pp. 3/1–312, doi: 10.1049/ic:20000082.

[9] R. J. Plowman, 'Wires and plates an introduction to grounding and shielding'. IEE Seminar on Shielding and Grounding (Ref. No. 2000/016), London, UK, 2000, pp. 1/1–110, doi: 10.1049/ic:20000080.

[10] P. N. Micropoulos and T. E. Tsovilis, 'Lightning attachment models and perfect shielding angle of transmission lines'. 44th Universities Power Engineering Conference, Glasgow, Scotland, Paper number 57-70151, 2009.

[11] R. Hileman, Shielding of transmission lines. Insulation Coordination of Power Systems. CRC Press, Taylor & Francis Group, New York, NY, USA; 1999, pp. 1–6.

Chapter 8
Lightning and surge protection

This chapter deals with lightning protection practices followed in overhead lines. After an introduction, it describes lightning strokes, its formation and characteristics. Return lightning discharge or return stroke and multiple strokes have been discussed. Then, different protection schemes against lightning have been discussed. Earth wire or sky wire has been described. Different types of lightning arresters like rod gap, horn gap, sphere gap, multiple gap, impulse type, valve type, expulsion type, auto valve type, etc. have been described. Wave propagation in transmission line has been explained with the help of characteristic impedance, propagation constant, etc. Different overvoltage assessment methods have been described related to lightning protection on overhead lines.

8.1 Introduction

Overhead lines are exposed to open air and therefore, they are subject to overvoltages due to both external and internal causes. Lightning is the main cause of the occurrence of overvoltage in overhead lines. Different types of transients are experienced by overhead lines [1–3]. Transients faced by lines may be divided into two following broad categories:

1. DC transients
2. AC transients

Lightning is an example of DC transients. AC transients are again may be divided into two categories as follows:

1. High frequency (it is seen for one or two cycles of multiple number of cycles)
2. Low frequency

The causes of transients may be written as follows:

1. Internal causes
   (i) Switching surges
   (ii) Energization of lines
   (iii) Sudden de-energization of opening of big load unit

2. External causes
   (i) Lightning
   (ii) Induced by external circuit

8.2 Lightning strokes

Lightning is unidirectional voltage waveform of very high peak. Both direct and indirect lightning strokes have severe adverse effect on transmission lines. It can cause breaking of insulators, lines and other members hampering the service of the electric power system at large. Therefore, it has drawn attention to take sufficient during design, installation, erection, operation and also maintenance for saving the system from adverse lightning effects.

8.3 Formation

Origin of the lightning starts from the causes of formation of clouds, their various physical properties and wind-driven movement due to difference in air pressure. Clouds get charged during their movement and collision. It exposes itself by sound and discharge through air to the earth. Human beings do not have any direct control over the occurrence and severity of lightning. To design lightning proof overhead lines and provide protection for the power system network connected with the line against lightning, detailed study on the characteristics of lightning is needed.

Lightning takes place in following three steps:

- Accumulation of charge
- Formation of streamer
- Lightning discharge

8.3.1 Accumulation of charge

In the formation of cloud, when air moves up, temperature decreases up to −25 °C. Moisture present in the air forms very small tiny water droplets and relatively bigger graupel. In collision among air and water droplets of different sizes, they get charged. Tiny droplets being positively charged go up. Relatively bigger graupel being negatively charged goes down. Thus, upper part of cloud becomes positive and lower part becomes negative. This develops potential difference between them. Also, the clouds are separated from earth which is considered as a conductor; the separating medium air is a dielectric medium. Negatively charged clouds induce positive charges in earth surface and potential difference is developed. However, it should be remembered that the distribution of charged air droplets also depends on some other factors like wind flow, air movement in particular region, temperature difference, and pressure difference.

Polarity of charges in cloud: It may be positive or negative. However, negative charge formation at the lower part of cloud is observed in more than 80% cases.

8.3.2 Formation of streamer

After raising charge concentration, when developed potential difference becomes higher than the limit of dielectric property, charge concentration starts frame streaming of lightning and moving towards the centre of another charge located in different parts of same cloud or different cloud or towards earth.

The path of movement of concentrated charges is known as a streamer. The tip of this streamer is known as a leader. Other parts of the streamer move in the direction opposite to the leader. Based on the speed of the leader, streamer can be divided into two broad categories as follows:

1. Very fast streamer
2. Relatively slow streamer

Direction of leader may be downward, upwards or in other direction. There are different opinions on the location of streamers. Some opine that it occurs in the interface between hot and cold air, i.e., when air layers of different temperature get in touch each other.

8.3.3 Lightning discharge

It is electrostatic discharge occurred within intra-cloud (IC) or between two clouds or between cloud and earth having charge distribution of different polarity. When charged streamer touches earth or other clouds or other parts of the same cloud of opposite polarity, it is known as strikes. Sometimes, another streamer is formed from the side of earth of other cloud sections. Both steamers move towards each other and touches. Based on streamer movement, lightning discharge may broadly be divided into following three categories:

1. Inter-cloud (IC) lightning discharge: It occurs between different opposite charge centres of the same cloud.
2. Cloud to cloud (CC) discharge: It occurs between clouds of opposite charge centre.
3. Cloud to ground (CG) lightning discharge: It occurs between cloud and earth of oppositely induced charge surface.

Two types of cloud to ground (CG) lightning discharges are seen as follows:

1. Negative lightning discharge: Discharge from negatively charged cloud to positive ground surface.
2. Positive lightning discharge: Discharge from positively charged cloud to negative ground surface.

Lightning is associated with ionization of air, plasma layer. During this process, heat is generated and heats up the air. Then, lightning-driven movement of air starts, resulting in sound. Thus, thunder is formed. Two categories of streamers (direct and return) have been shown in Figure 8.1.

The formation of charge has been provided by two well-accepted explanations: Wilson's theory and Simpson's theory. Based on different shapes of streamer, lightning discharges have been given many different names like:

1. Spider or anvil lightning
2. Ball lightning
3. Bead lightning
4. Cloud to air lightning
5. Dry lightning
6. Forked lightning
7. Heat lightning
8. Ribbon lightning
9. Sheet lightning
10. Smooth channel lightning
11. Staccato lightning
12. Superbolts lightning
13. Sympathetic lightning
14. Upwards lightning
15. Clear air lightning

8.4 Characteristics

Lightning is unidirectional. It rises sharply with very short period of time and reaches peak value. Then, it falls for longer period of time. Decaying part is known as a tail. A typical waveform of lightning has been shown in Figure 8.2.

The waveform has following three main parts:

- Front: It refers to rising part of the wave. It lasts for up to 10 μs.
- Crest: It refers to the peak of the waveform; it can reach up to 450 kA.
- Tail: It refers to the decaying part of the waveform. It can last up to period of 10–100 μs to 1,000 μs to fully die out.

Characteristics of lightning wave can be mathematically modelled by a combination of two exponential functions as follows:

$$v = V\left(e^{-k_1 t} + e^{-k_2 t}\right) \tag{8.1}$$

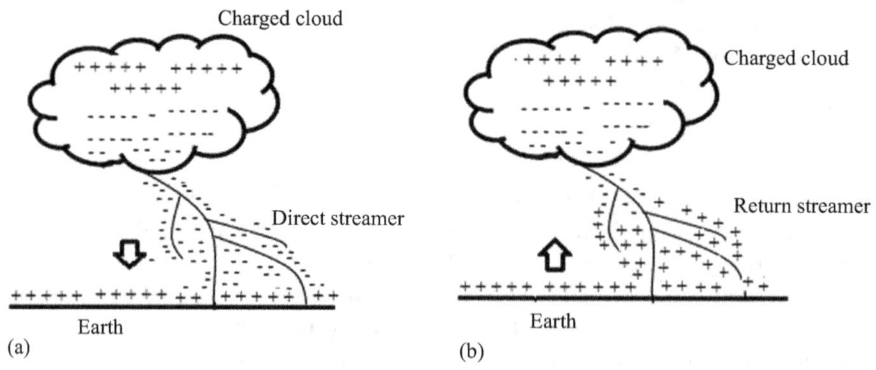

Figure 8.1 Lightning formation: (a) direct streamer (b) return streamer

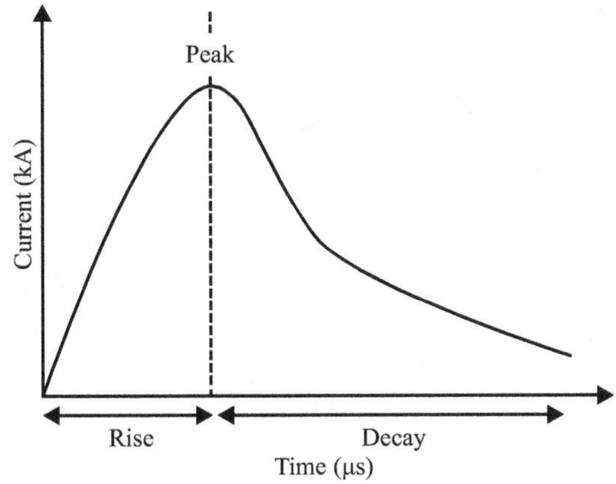

Figure 8.2 Lightning characteristics

The shape will be determined by two constants: k_1 and k_2.

Lightning wave is characterized by the following main three parameters:

- Peak value: It refers to the maximum value of the wave.
- Rise time: It refers to the time required to reach peak value.
- Decay time: It refers to the time after the occurrence of peak to reduce up to 50% of its peak value.

If R is the resistance in the path of lightning stroke through tower (include tower footing), the voltage (V) can be expressed as follows:

$$V = RI \tag{8.2}$$

Considering inductance across the tower as L, induced voltage may be written as follows:

$$V_L = L_T \frac{di}{dt} \tag{8.3}$$

where, L_T is the inductance across the tower.

This voltage will relate to the electric field induced due to lightning. Electric field can be written as follows:

$$E = \frac{V_L}{h} \tag{8.4}$$

where, h is the height of the tower.

8.5 Return lightning discharge or return stroke

During lightning, leader of streamers starts from lower part of the cloud, which is negatively charged. In the path of streamer, air breakdown occurs and air path

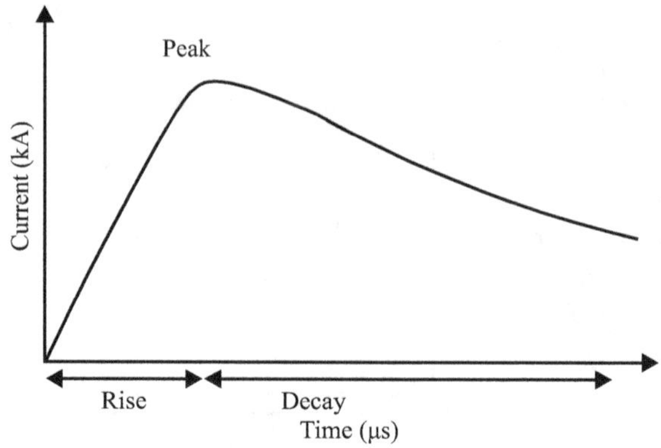

Figure 8.3 Lightning characteristics for return stroke

becomes conductive in nature. When negative streamer reaches earth, another streamer of positive nature develops from earth side and goes up. It attempts to neutralize the cloud charge. Streamer moving towards cloud and having opposite polarity of the main streamer is known as return lightning discharge or return strokes. It can speed up to one-tenth of speed of the light, can sustain up to 100 μs and causes severe overvoltage. Breakdown occurs during discharge of avalanche type and may cause severe damage in the surround lines or infrastructure. Return path may include trees, structure, towers, etc. Sample characteristic of return strike has been shown in Figure 8.3; however, it may be different for different return strokes. It may be noted that streamer velocity in return strokes is much less than first direct stroke and it sustains longer period of time. Sometimes, oscillatory nature is also observed and current may have both negative and positive magnitudes at different instances.

8.6 Multiple strokes

Lightning strokes are accompanied by return as well as several other strokes that can occur in inter-cloud (IC) regions. When return goes up and touches cloud, that part of the cloud either neutralizes or changes its own polarity. This may form sufficient potential difference with other parts (mainly upper part) of the same cloud or other clouds and generates more discharges. These are referred to as multiple strokes.

8.7 Frequency and intensity

Several lightning strokes can take place within a small interval of time. Time interval between strikes may be up to 0.0005 s. Intensity of the strikes gradually decreases. Human being does not have direct control over frequency and intensity of lightning. However, statics show that frequency is increasing with the increase of pollution level of air and UV interference with climatic condition.

If we consider the frequency of occurrence of lightning in a region, it depends on the climatic condition of the region. Some regions in the earth are seen to have experience of lightning throughout the year.

8.8 Effect of lightning and protective measures

Lightning has many adverse effects in the surrounding atmosphere and overhead lines. Effects on atmosphere include

- Formation of air breakdown in the path
- Heating up of the air, causing its difference in temperature and pressure followed by thunderstorm
- Formation of plasma
- Development of high peak current and voltage
- Heating up of the trees, infrastructure, etc. located in the path or vicinity of the stroke

It has many adverse effects on an unprotected overhead system as follows:

- Direct stroke causes flow of high current and high voltage.
- High voltage due to lightning can damage insulators if no protective measures are taken.
- It can damage support structure if they are not properly grounded.
- High voltage occurred due to lightning propagates to the substation, generating station and other utility equipment and causes severe damage if they are not protected against lightning and overvoltages.

Protective measures taken against lightning include protection taken in substation, generating station and utility system as well as protection in overhead lines. Protections taken against lightning are as follows:

- Ground of supports used in distribution and transmission lines.
- Use of earth wire in 110 kV lines or above.
- Installation of lightning arrester or surge diverter.
- Neutral grounding in low-voltage distribution system by different means wherever applicable.

8.9 Earthing

For safety and protection against lightning, grounding is must in overhead transmission lines. Whether it is a distribution line or transmission line, earthing of line support must be incorporated as per guidelines and standard. From the substation end, neutral must be grounded. Each support of at least interval of four supports must have proper earthing. In transmission lines, all the four legs or one out of four legs must be grounded and the earthed leg should be specifically mentioned by the designers. For river crossing towers, at least two diagonally located legs must be grounded. Support end of insulator opposite to the line conductors, guard wire, etc. must be grounded. Separate earth wire along the post or sometimes structure may be used for earthing path.

8.10 Earth wire or sky wire

Earth wire is used at top (as shown in Figure 8.4) of the lattice tower of transmission lines directly with the metal body. One, two or all legs are grounded. Earth wire protects the line from lightning strokes. It attracts charges and discharge to ground. As earth wire is positioned at the top, it is also called sky wire. At normal operating time, current is practically zero through earth wire when there is no lightning discharge. It also helps to protect from various earth faults by passing earth fault currents to ground. The number of earth wire depends on voltage level of the overhead line and height of the tower. Earth wire is positioned as sufficiently higher position than phase wires and care should be taken to check the sag of earth wire, and it should be such that at mid-span of the line, sufficient gap is present between earth wire and phase wire. For overhead lines of 110 kV and above, earth wire is a must to be included in design and installation for the safety of the line, networks and the population passing under the line.

Normal composite wire may be used as earth wire. In some application, optical ground wires have been introduced where optical fibre has been reinforced in the core. This fibre helps in transferring communication signal for generation purpose as well as for power-line communication. Fibre used in such application should have very high temperature withstanding capability.

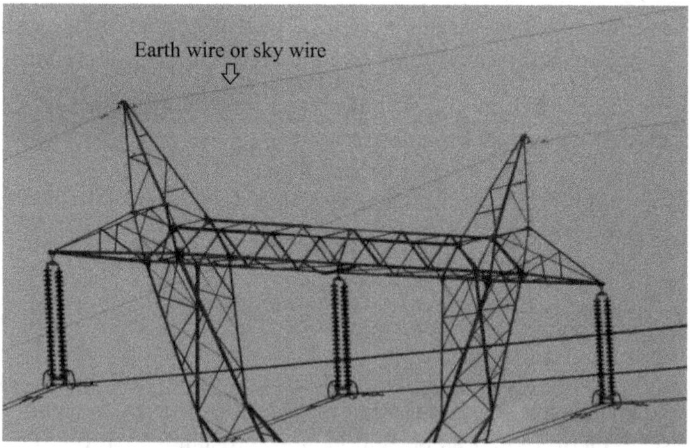

Figure 8.4 Earth wire

8.11 Shielding by earth wire

Earth wire attracts charges induced by clouds and passes them to the ground through earthing system. In this way, it shields phase wires from lightning. Shielding angle refers to the angle made by phase wire with the vertical line passing through earth wire. It may be positive or negative depending upon the relative position of earth wire and phase wires [3]. In Figure 8.5, positive and negative shield angles have been shown separately. Shield angle must be set during the design time based on the selection from process as per guidelines of the regulatory body where the transmission line is being installed. Normally, its value is chosen much less than 30°.

8.12 Surge impedance of earth wire

Earth wire also passes oscillatory current for various fault signals. It possesses a considerable amount of frequency-dependent reactance part. Therefore, instead of considering resistance only, impedance offered by earth to the fault current should be considered for better modelling and analysis. Surge impedance of earth line refers to the impedance offered by earth wire to fault current which consists of resistance as well as frequency-dependent reactance.

8.13 Other overvoltages

Other overvoltages can occur in power lines due to different switch and fault effects. These may broadly be divided into two categories:

- High-frequency overvoltage
- Low-frequency overvoltage

Causes of high-frequency overvoltage include

- Fault occurrence, circuit breaker operation and clearance of fault
- Isolation of transformer
- Switching effect of reactive power units
- Ferro resonance

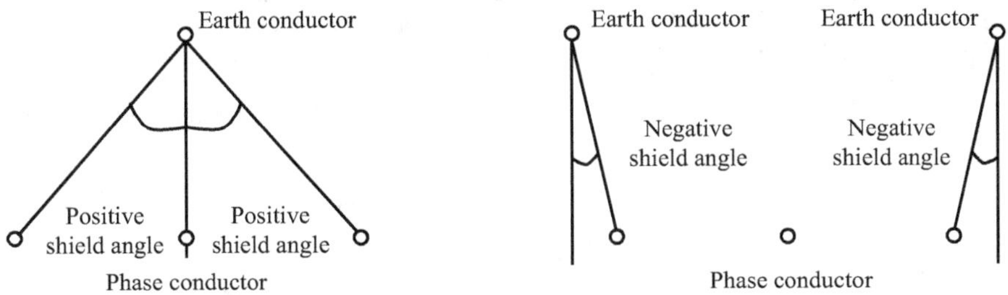

Figure 8.5 Shield angle

Causes of low-frequency overvoltage include

- Sudden change in load distribution
- Earth fault
- Ferranti effect

The above-mentioned causes of overvoltages other than lightning in overhead power lines have been presented in short in Table 8.1.

8.14 Line faults

Different probable line faults may be divided into two categories:

- Symmetrical fault
- Unsymmetrical fault

Symmetrical fault is of two categories:

- Symmetrical triple-line fault
- Symmetrical triple-line-to-ground fault

The unsymmetrical fault can occur in the following ways:

- Line-to-ground fault
- Line-to-line or double-line fault
- Line-to-line-to-ground or double-line-to-ground fault

The unsymmetrical fault is characterized by generation of negative sequence current component, whereas unsymmetrical ground fault is characterized by generation of zero sequence current component.

Different types of relays are used for protection against line fault. For example, earth fault relay, over current relay, negative sequence relay, directional current relay, etc.

Table 8.1 Causes of overvoltages

Causes	Description
Lightning	External cause
Fault occurrence, circuit breaker operation and clearance of fault	May generate from overhead lines, buses, etc. and their associated switchgear units
Isolation of transformer	It can take place at substation ends of overhead lines
Switching effect of reactive power units	It can take place at reactive power control stations
Ferro resonance	It can occur relative interaction of inductive and capacitive reactances
Sudden change in load distribution	It depends on load variation and its regulation
Earth fault	Faults in connection with earth
Ferranti effect	It is due to line capacitance in medium and long transmission lines at no or low load condition

Some special protection schemes have also been introduced like core balance relay.

Differential protection scheme has also been used in different forms.

Line protection scheme is characterized by two modes as follows:

- Current-graded protection scheme
- Time-graded protection scheme

The specific difference between line protection and device protection is that device protection area is localized. But as line covers wide area, protection scheme is expected to perform two things: detection of fault and detection of location of fault. For this purpose, different types of voltage-to-current-ratio-based distance protection schemes are used like impedance relay, admittance relay and reactance relay, resistance or ohm (R) relay (for DC). Distance relay provides protection zone wise along with back-up protection. Care is taken so that no zone remains unprotected; there may be overlapped protective zone.

8.15 Wave propagation in transmission line

Long transmission lines (LTLs) have very high value of both line inductance and capacitance. Resistance of LTL (higher than short and medium transmission lines) is small with respect to inductance and capacitance and hence is often neglected in modelling and analysis. All line parameters are distributed in nature.

8.15.1 Modelling

Modelling of LTL has been discussed in Chapter 2 which gives the relation between sending end and receiving end voltages using transmission line parameters as follows:

$$\begin{pmatrix} V_S \\ I_S \end{pmatrix} = \begin{pmatrix} \cosh \gamma x & Z_C \sinh \gamma x \\ \frac{1}{Z_C} \sinh \gamma x & \cosh \gamma x \end{pmatrix} \begin{pmatrix} V_R \\ I_R \end{pmatrix} \tag{8.5}$$

Transmission line parameters are

$$\begin{pmatrix} A & B \\ C & D \end{pmatrix} = \begin{pmatrix} \cosh \gamma x & Z_C \sinh \gamma x \\ \frac{1}{Z_C} \sinh \gamma x & \cosh \gamma x \end{pmatrix} \tag{8.6}$$

where Z is the line impedance per unit length,

Y is the line admittance per unit length and x is the line length.

$$\sqrt{YZ} = \gamma \text{ and } Z_C = \sqrt{\frac{Z}{Y}}$$

Condition of symmetry can be judged by the relation: $A = D = \cosh \gamma x$

Condition of reciprocity can be judged by the relation: $AD - BC = 1$

LTL is symmetrical as well as reciprocal in nature.

8.15.2 Characteristic impedance

Transmission line is characterized by characteristics impedance. It is determined by the square root of open-circuit impedance and short-circuit impedance of any side. Using transmission line parameters, open-circuit impedance (Z_{OC}) measured at sending end when receiving end is open can be written as

$$Z_{OC} = \frac{A}{C} = \frac{\cosh \gamma x}{\frac{1}{Z_C} \sinh \gamma x} = \frac{Z_C \cosh \gamma x}{\sinh \gamma x} \tag{8.7}$$

Similarly, using transmission line parameters, short-circuit impedance measured at sending end when receiving end is short can be written as

$$Z_{SH} = \frac{B}{D} = \frac{Z_C \sinh \gamma x}{\cosh \gamma x} \tag{8.8}$$

Therefore,

$$\sqrt{Z_{OC} Z_{SH}} = \sqrt{\frac{AB}{CD}} = \sqrt{\frac{Z_C \cosh \gamma x}{\sinh \gamma x} \times \frac{Z_C \sinh \gamma x}{\cosh \gamma x}} = Z_C \tag{8.9}$$

Thus, Z_C can also be determined by the following relation:

$$Z_C = \sqrt{\frac{Z}{Y}}$$

8.15.3 Wave propagation

When line carries electric power, the line and its surroundings are associated with electric field intensity (E) and magnetic field intensity (H); together they form electromagnetic waves. The resultant of two fields can be expressed by the cross product $(E \times H)$ which is commonly known as a pointing vector. Resultant wave also propagates. Propagation of this resultant is characterized by characteristic impedance and propagation constant. When wave propagates from one medium characterized by characteristic impedance to other medium characterized by some other characteristic impedance, some parts of the wave get reflected and other parts get transmitted.

8.15.4 Propagation constant

Propagation of wave is decided by propagation constant. $\gamma(= \sqrt{YZ})$ is called propagation constant of the line. Using shunt admittance and series impedance per km of the transmission line, this can be determined.

8.15.5 Image impedance

The transmission line delivers maximum power to the load when complex conjugate of load impedance equals the characteristic impedance. At the receiving end, impedances of both sides are equal. Thus, characteristics impedance may be considered as image of the load impedance and vice versa. At this condition, characteristics impedance is called image impedance. For propagation wave, no reflection is observed at this condition.

8.15.6 Image impedance loading

When characteristic impedance matches with load impedance, i.e., load impedance is complex conjugate of characteristics impedance, the transmission line will transfer maximum power to the load. The maximum power delivered during this condition is known as an image impedance loading. Mathematically,

$$\text{Image Impedance Loading (IIL)} = V_R I_R = \frac{V_R^2}{Z_C} \tag{8.10}$$

8.15.7 Velocity and wavelength of propagation wave

Propagation constant can be expressed in complex mathematical form by attenuation constant (α) and phase constant (β) as:

$$\gamma = \alpha + j\beta \tag{8.11}$$

In terms of line parameters, it can be written as:

$$\gamma = \sqrt{YZ} = \sqrt{(r + j\omega L)(g + j\omega C)} \tag{8.12}$$

For lossless line, real part of impedance (r) and real part of admittance (g) are absent. Thus,

$$r = 0$$
$$g = 0$$

Therefore, propagation constant will consist of only imaginary part as follows:

$$\gamma = j\omega\sqrt{LC}$$
$$\therefore \gamma = j\beta \text{ (where, } \beta = \omega\sqrt{LC})$$

In lossless line, velocity of propagation wave is given as follows:

$$v = \frac{\omega}{\beta} = \frac{1}{\sqrt{LC}} == \frac{1}{\sqrt{\mu_0 \varepsilon_0}} \tag{8.13}$$

In lossless line, wavelength (λ) of propagation wave can be given as follows:

$$\lambda = \frac{2\pi}{\beta} = \frac{2\pi}{\omega\sqrt{LC}} = \frac{1}{f\sqrt{LC}} = \frac{1}{f\sqrt{\mu_0 \varepsilon_0}} \tag{8.14}$$

8.15.8 Wave reflection and standing wave

If wave propagates from one medium characterized by characteristic impedance to other medium characterized by some other characteristics impedance, some parts of the wave get reflected and other parts get transmitted. Therefore, at the interface of two different medium of wave propagation, incident wave is divided into two parts: transmitted wave and reflected wave.

Reflection coefficient is defined by the ratio of amplitude of reflected wave to amplitude of incident wave. Mathematically, in terms of characteristic impedance (Z_C) and load impedance (Z_L), it can be expressed as

$$\text{Reflection coefficient } (\Gamma) = \frac{Z_L - Z_C}{Z_L + Z_C} \tag{8.15}$$

Reflection coefficient ranges from 0 to 1. When magnitude of load impedance ($\boldsymbol{Z_L}$) equals magnitude of characteristic impedance ($\boldsymbol{Z_C}$), it will become

$$\text{Reflection coefficient } (\Gamma) = \frac{Z_L - Z_C}{Z_L + Z_C} = \frac{Z_L - Z_C}{Z_L + Z_C} = 0 \tag{8.16}$$

It indicates that, at this condition, there will not be any reflection.

Reflected wave goes back to the incident medium. Resultant of incident wave and reflected wave is known as a standing wave. A standing wave is characterized by periodic rise and fall of its amplitude.

The ratio of maximum amplitude to minimum amplitude is referred to as standing wave ratio (SWR).

$$SWR = \frac{1 - |\Gamma|}{1 + |\Gamma|} \tag{8.17}$$

For no reflection,

$$SWR = \frac{1 - |\Gamma|}{1 + |\Gamma|} = 1 \tag{8.18}$$

8.15.9 Protection against travelling waves

Travelling wave propagated through the line causes damage to different parts of the system. For the protection of lightning and other overvoltages, insulation coordination and proper grounding are important [4,5]. Direct lightning strokes are protected by earth wire and proper ground system. But travelling waves and other overvoltages of high peaks are protected by surge arresters. They provide high level of insulation at normal condition. But when peak of waves crosses the limit of arresters, they are bypassed by discharging to the ground. Different features and types of lightning arresters are discussed in the following sections.

8.16 Surge arresters

Surge arresters are popular device widely used to discharge surges and give protection against various types of surges. Some commonly used surge arresters are as follows:

- Rod-gap surge arrester
- Horn-gap lightning arrester

- Sphere-gap lightning arrester
- Multiple-gap lighting arrester
- Impulse-type lightning arrester
- Valve-type lightning arrester
- Expulsion-type lightning arrester
- Auto-valve-type lightning arrester
- Thyrite lightning arrester
- Metal oxide lightning arrester

Different types of lightning arresters used have been discussed in the following subsections. Major specification or properties of lightning arresters are as follows:

- Rated current
- Follow current
- Discharge current
- Nominal discharge current
- Maximum discharge current
- Rated voltage
- Power frequency spark overvoltage
- Impulse spark overvoltage
- Discharge voltage
- Coefficient of earthing
- Non-linear resistor

8.16.1 Rod-gap lightning arrester

Rod-gap lightning arrester is the simplest type of lightning arrester. It consists of two rods facing each other separated by air. Rods are bent at right angles. One rod is connected to line conductor and another rod is connected to the support end of insulator which is connected with earth. Tips of the rods are made pin type facing each other so that all energy discharge is directed in particular direction. Rod-gap lightning arrester has been shown in Figure 8.6.

Normal operating voltage of the line does not cause discharge. When high impulsive voltage appears in the line, it is diverted to earth by discharge taken place through gap. Gap between two rod-type electrodes is designed in such a way that break voltage required to cause discharge is higher than the operating line voltage and at least around 20–25% less than the high impulsive voltage against which it is expected to provide protection. Discharge may damage the insulator surface area; to minimize the chance of damage, minimum distance between the insulator and rod-gap discharge path of around one-third of the length of gap is maintained.

8.16.2 Horn-gap lightning arrester

Horn-gap lightning arrester consists of two metal electrodes separated by air and placed in such a way that looks like a horn. One electrode is connected to line conductor and another electrode is connected to the support end of insulator which is connected with earth. Sometimes, line-end electrode is connected to a high resistance to the line. Horn-gap lightning arrester has been shown in Figure 8.7.

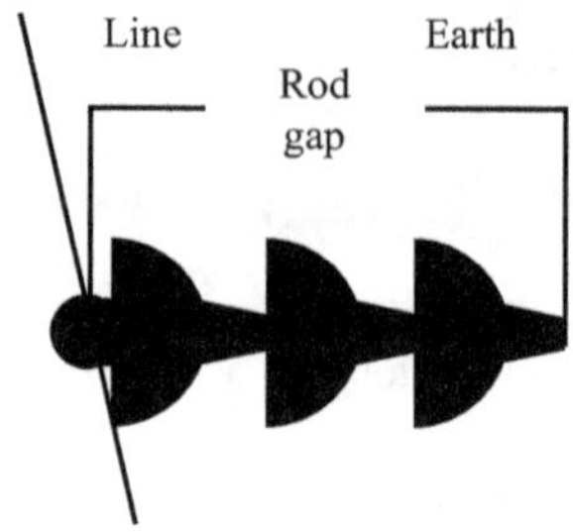

Figure 8.6 Rod-gap lightning arrester

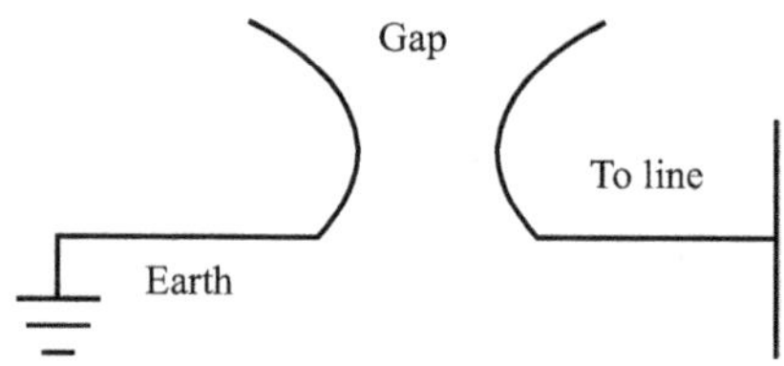

Figure 8.7 Horn-gap lightning arrester

The operating principle of horn-gap lightning arrester is almost same as other types of lightning arrester. Normal operating voltage of the line does not cause discharge; but when high impulsive voltage appears in the line, it is diverted to earth by discharge taken place through gap. Gap between two horn-type electrodes is designed in such a way that break voltage required to cause discharge is higher than the operating line voltage and at least around 20–25% less than the high impulsive voltage against which it is expected to provide protection. Discharge may damage the insulator surface area. To avoid or minimize damage, minimum distance between the insulator and horn-gap discharge path is maintained.

8.16.3 Sphere-gap lightning arrester

Sphere-gap lightning arrester consists of two metal electrodes of spherical shape separated by air. Like rod-gap or horn-gap-type lightning arrester, one electrode is connected to line conductor and other electrode is connected to the support end of insulator which is connected with earth. Sometimes, line-end electrode is connected a high resistance to the line. Sphere-gap lightning arrester has been shown in Figure 8.8.

At normal operating voltage of the line, discharge does not take place between two spheres; but when high impulsive voltage appears in the line, discharge takes place through the gap. Sometimes, a high resistance is added in the path of electrode and line.

8.16.4 Multiple-gap lightning arrester

In multiple-gap lightning arrester, instead of one gap, multiple number of gaps are created between two extreme-end electrodes. Multiple-gap lightning arrester has been shown in Figure 8.9.

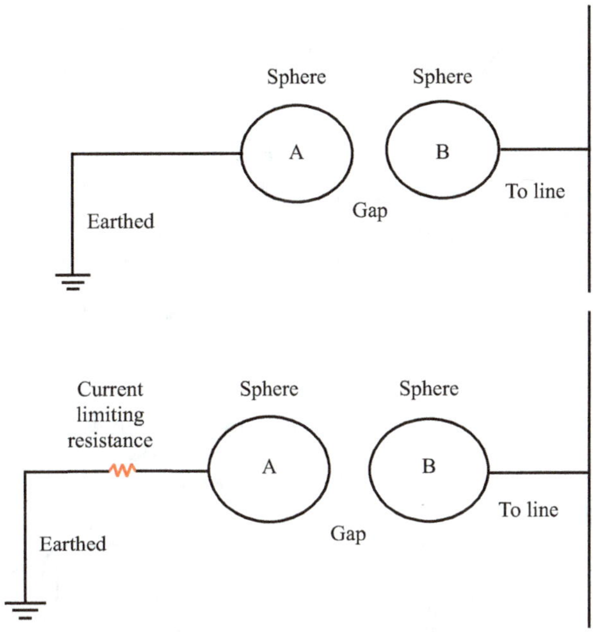

Figure 8.8 Sphere-gap lightning arrester

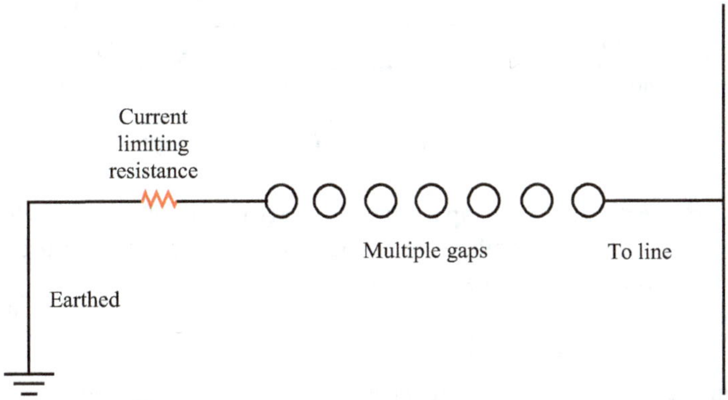

Figure 8.9 Multiple-gap lightning arrester

8.16.5 Impulse-type lightning arrester

Impulse-type lightning arrester has been shown in Figure 8.10. An additional notch-type auxiliary electrode is placed in between two horn electrodes. A resistance is connected in series with the gap.

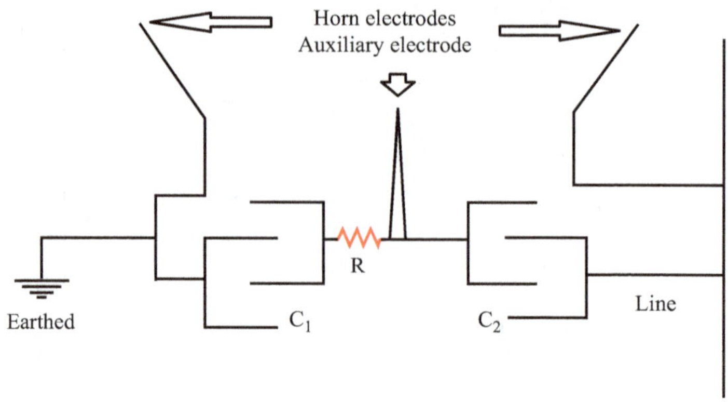

Figure 8.10 Impulse-type lightning arrester

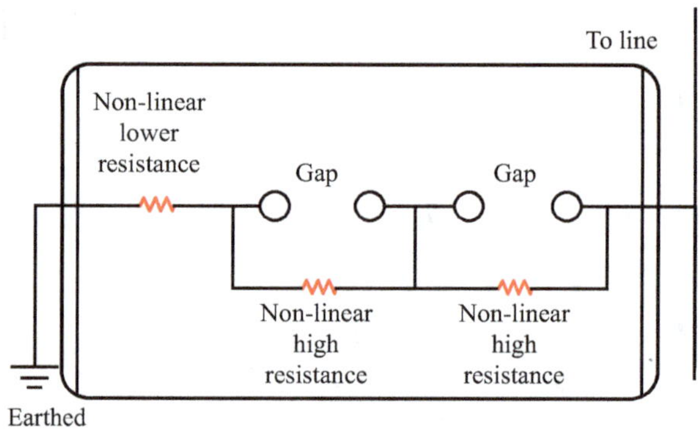

Figure 8.11 Valve-type lightning arrester

Auxiliary electrode forms capacitances with two horn electrodes which can be adjusted to get control over surge diversion. Outermost electrodes are connected with line and earth in two opposite sides.

8.16.6 Valve-type lightning arrester

In valve-type lightning arrester, multiple gaps are provided. In addition to this, high non-linear resistances are provided across the gaps and in the path of electrode and line in series. Valve-type lightning arrester has been shown in Figure 8.11. At normal operating voltage, discharge does not take place. High resistance ensures no current. The whole arrangement of gaps and resistance is enclosed by porcelain materials. But when high voltage appears in the line, discharge takes place and high voltage is diverted to ground.

8.16.7 Expulsion-type lightning arrester

Expulsion-type lightning arrester consists of two gap arrangement: one is made inside fibre tube and the other is in series with the first external to the tube. One

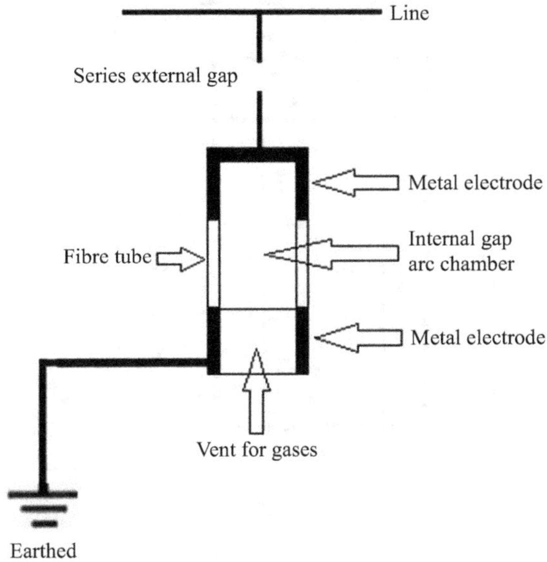

Figure 8.12 Expulsion-type lightning arrester

electrode is made hollow in shape. Expulsion-type lightning arrester has been shown in Figure 8.12. At normal operating voltage, discharge does not take place. But when high voltage appears in the line, discharge takes place and high voltage is diverted to ground. During discharge generated, gas goes out through the hollow electrode side.

8.16.8 Auto-valve-type lightning arrester

This is special valve-type lightning arrester which is more robust in operation. Mica is used as insulating material in addition to porcelain. Mica rings help in smooth operation at normal voltage by providing proper insulation. This can be effectively used in overhead lines ending at automated substation and at isolated substation. Its manufacturing and maintenance costs are comparatively less than other valve-type or metal–oxide-type arresters.

8.16.9 Metal–oxide-type lightning arrester

In metal-oxide lightning arrester, multiple gaps are replaced by metal-oxide. Resistances are made of zinc oxide. It helps stable operation of surge diversion. Zinc oxide is N-type in nature. Different types of oxide impurities are added with zinc oxide like oxides of antimony, chromium, cobalt, manganese, etc.

Zinc oxide with impurity is dried and is formed. They are coated with conducting material. The whole arrangement is enclosed by silicone rubber or sometimes by porcelain. Metal-oxide lightning arrester has been shown in Figure 8.13. By the property of semiconductor material with impurity, potential barrier is

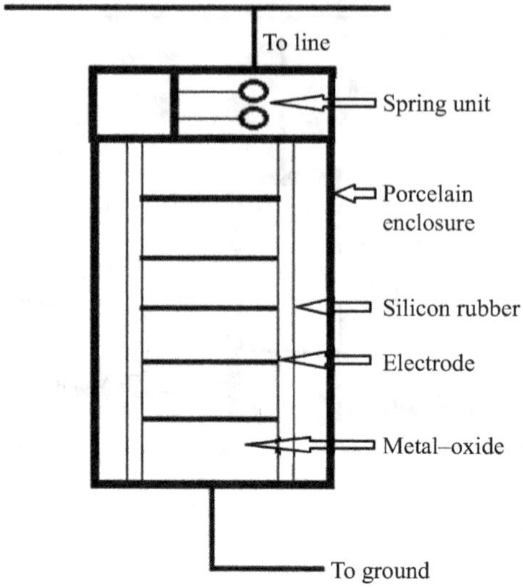

Figure 8.13 Metal oxide lightning arrester

developed across zinc oxide grains. This prevents flow of current. Almost zero current flows at normal voltage. When high voltage appears in the line, breakdown occurs, current flows and surge is diverted to ground. Metal oxide lighting arresters are used both in AC and DC overhead lines. Some advantages of metal-oxide-type lightning arrester are as follows:

- Stable operation
- Very low leakage current
- No power flow current
- High capacity
- Less damage possibility
- Control of surge current is possible

8.16.10 Thyrite-type lightning arrester

Thyrite-type lightning arrester is used for diverting very high voltages. Multiple numbers of disc separated by gap are placed in stack and enclosed in porcelain container. Gap reduces flow of current at normal condition. Above stack of disc, thyrite disc is place. Above thyrite disc and at bottom of porcelain container, aluminium alloy is provided. Top is designed with cap and metal grip to hold line conductor through clamp. At bottom, metal grip is provided to hold ground wire through clamp for earthing. Thyrite-type lightning arrester has been shown in Figure 8.14. At normal operating voltage, discharge does not take place. Thyrite provides high resistance in the path of surge current. But when high voltage appears in the line, discharge takes place and high voltage is diverted to ground.

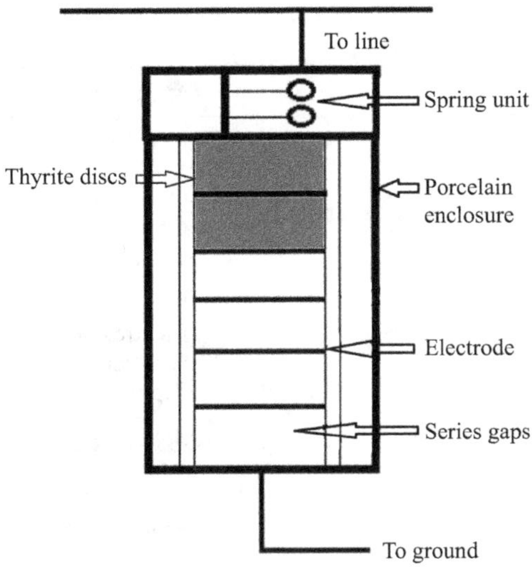

Figure 8.14 Thyrite-type lightning arrester

8.17 Surge absorber

Surge absorber absorbs energy from travelling waves and reduces their high amplitude. It works on corona discharge phenomena. Surge absorber is connected in series with the line and placed near the load terminal. Ferranti surge absorber is one common-type surge absorber. It consists of an inductor which is connected in series with the line. The inductor is placed inside a closed cylindrical metal container. The metal enclosure is grounded properly. Travelling wave passes through the inductor. If the magnitude of travelling wave crosses critical corona value, energy is passed to ground through the container reducing the amplitude of travelling wave. The equivalent of this surge absorber may be considered as transformer having primary number of turns equals the number of turns of the inductor and secondary of single turn contributed by the closed metal container which is grounded.

8.18 Overvoltage measurement

Overvoltage measurement should be carefully carried out taking safety measures and concerns on accuracy and approximation. Standards and guidelines are to be followed strictly in carrying out measurement of high voltages in laboratory as well as in live line. Some overvoltage measurement methods are given in Table 8.2.

8.18.1 Sphere gap

Two spherical electrodes are taken and voltage is applied across the spheres. Discharge is allowed to take place. Discharge between two spheres depends on the

Table 8.2 Different methods for measurement of high voltages

High-voltage signals	Methods
Alternating	Using sphere gaps Voltage divider by capacitor Capacitor-based conversion to current Capacitor voltage transformer Digital recorder
Direct or unidirectional	Sphere gap Electrostatic voltmeter Generating voltmeter Ripple meter
Impulse	Sphere gap Voltage divider using capacitor Voltage divider using resistors Delay cables Digital approaches

breakdown voltage of the medium and distance between them. It is normally expressed in terms of kV/mm. Voltage across the sphere is measured by

$$\text{Voltage peak} = \text{Breakdown voltage} \times \text{gap between spheres}$$

The accuracy of the measurement techniques depends on other various factors like diameter of the sphere, density of the dielectric medium, humidity, etc. Spheres are mounted at specified height and are made of specified diameter based on the range of the voltage to be measured. With the increase of measured parameter, diameter is made larger. As dielectric medium, air is widely used for high-voltage measurement. Density, humidity, pressure and temperature are dependent factors. To include their effects, measured value is multiplied by multiplying factor to get the actual value. Sphere-gap-based measuring arrangement has been shown in Figure 8.15. This method can be applied for measurement of alternating, direct or impulse voltages.

Discharge current is equivalent to short-circuit current which can damage the source of voltage considered for measurement. Before discharge, current is almost zero. During discharge, current suddenly rises to high magnitude. This can be limited by inserting resistance in series with the source. Relay arrangement is also sometimes incorporated to break the circuit just after discharge. Maximum voltage magnitude is monitored and stored in memory unit and remains displayed.

8.18.2 Capacitor-based voltage divider

A series of capacitors is taken with high voltage withstanding capability. High voltage to be measured is applied across the series of capacitors. The applied voltage is divided among all capacitors connected in the series combination based on their capacitance value. Voltage across one capacitor is measured and the applied voltage is calculated. The capacitor-based arrangement has been shown in Figure 8.16.

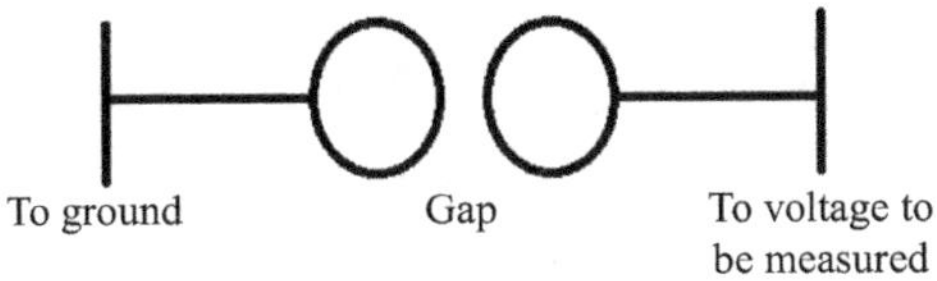

Figure 8.15 Sphere-gap-based high-voltage measurement

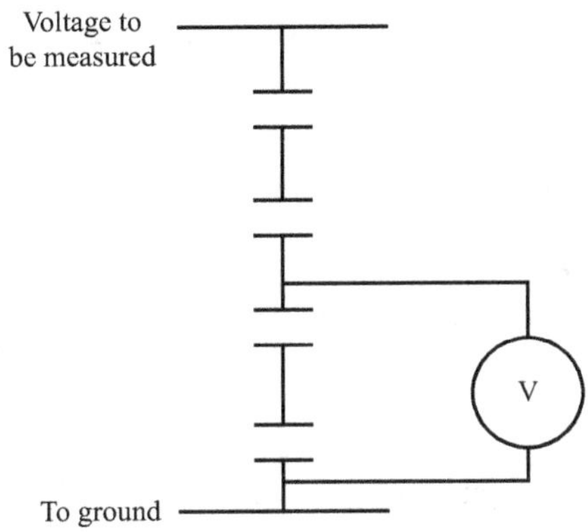

Figure 8.16 Capacitor-based arrangement

8.18.3 Voltage-converted current measurement

Voltage to be measured is applied across series capacitor and resistance arrangement. Capacitive current is divided into positive and negative cycles using two antiparallel diodes. Current in positive cycle flows through one diode and in negative cycle flows through another diode. Therefore, individual diode current gives current of one cycle. It is measured and applied voltage is calculated.

8.18.4 Resistance-based voltage divider

A series of resistance is taken with high voltage withstanding capability. High voltage to be measured is applied across the series of resistances. The applied voltage is divided among all resistances connected in the series combination based on their resistance value. Voltage across one resistance is measured and the applied voltage is calculated. The resistance-based arrangement has been shown in Figure 8.17.

8.18.5 Voltage-converted frequency-based digital measurement

Voltage to be measured is applied across series capacitor or a series of capacitors and resistance arrangement. Voltage is divided into components of series combination. Voltage across one capacitor or resistance is converted into frequency. Frequency component is analysed and measured in digital techniques.

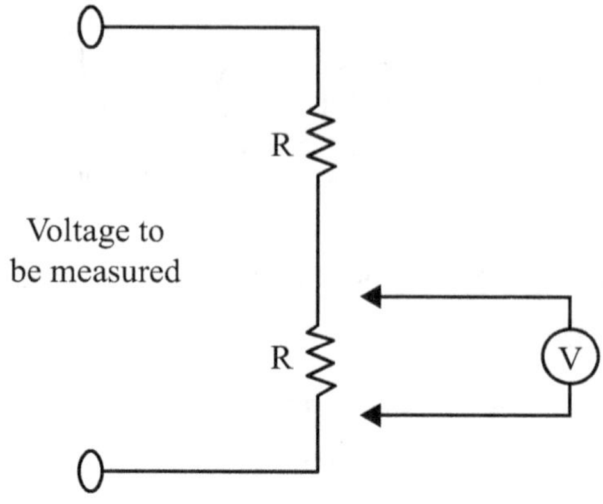

Figure 8.17 Resistance-based arrangement

Sometimes capacitive current is used to generate corresponding frequency signal for measurement. Capacitive current is divided into positive and negative cycles using two antiparallel diodes. Current in positive cycle flows through one diode and in negative cycle flows through another diode. Therefore, individual diode current gives current of one cycle. Current is passed through resistance and converted into frequency signals. It is measured and applied voltage is calculated.

8.18.6 Capacitor-based voltage transformer

The capacitor-based voltage transformer is used for measurement of high voltage. Voltage to be measured is applied across primary side. It reduces the primary voltage by capacitive divider method. Reduced secondary voltage is measured and primary applied voltage is calculated.

8.18.7 Digital recorder and impulse measurement

High voltage to be measured is scaled down by any of the above-mentioned methods. Reduced voltage signal is captured by data acquisition system. It is first passed through sampling and hold circuit. Each hold signal is converted into digital value by analogue-to-digital converter (ADC) and stored in a storing device. Schematic diagram of this method has been shown in Figure 8.18. It is a very convenient method for measurement of impulse which occurs for very short duration of time.

8.18.8 Electrostatic voltmeter

High voltage to be measured is scaled down by any of above-mentioned methods. Scaled-down voltage is measured by electrostatic voltmeter. High voltage is calculated accordingly.

8.18.9 Delay cable

High voltage to be measured is scaled down by any of the above-mentioned methods. Scaled-down voltage is passed through a resistive cable that adds delay

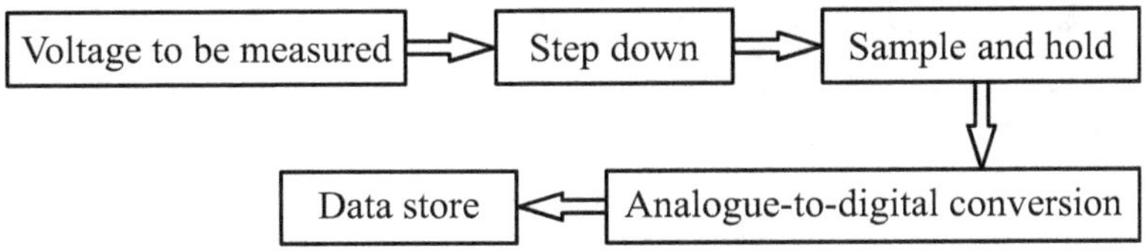

Figure 8.18 Schematic diagram for digital recorder and impulse measurement

with the signal without changing the shape. Then, the signal is captured through sample and hold circuit and an analogue-to-digital converter. Signals are stored in memory and analysed for measurement. This method is very helpful for measurement of high-voltage transients, which occurs for very short period of time.

8.19 Measurement of dissipation factor

Dissipation factor is the measure of dielectric strength of insulating medium. Smaller the value of dissipation factor, better is the dielectric strength of the medium or material. Dissipation factor is measured using bridge circuit. Schering bridge is commonly used for measurement of dissipation factor.

8.20 Measurement of partial discharge

High voltage is applied across the test material directly or through some filter. Gradually partial discharge is allowed through the test material. Test material is generally connected with an impedance in series. Partial discharge current is allowed to pass through it. Voltage across it is filtered, amplified and captured to analyse for measurement of partial discharge.

8.21 High-voltage testing

Different parts of overhead lines and connected devices undergo high-voltage test during installation, commissioning and maintenance. Some common high-voltage tests performed on insulators, lightning arrestors, transformers and bushings have been presented in Table 8.3.

8.22 Summary

Protection against lightning and other overvoltages is required for safe operation and to avoid damages that may occur due to high peak values of those signals. Phenomena of lightning have been discussed. Characteristics of direct streamer and return stroke have been explained. All these have been discussed in detail in this chapter. Natures of different overvoltages and wave propagation have been also discussed. Lightning has been protected by earth wires while other overvoltages are

Table 8.3 Different methods for high-voltage test

HV test object	Test
Line insulator	Power frequency voltage Impulse voltage Pollution
Lightning arrestor	Impulse current test
Transformer	Overvoltage test Impulse test: switching impulse, lightning impulse
Bushings	Impulse test

Table 8.4 Some useful standards and guidelines

Standards	Purpose
IEEE Std C37.30.3-2018	IEEE standard requirements for high-voltage interrupter switches, interrupters or interrupting aids used on or attached to switches rated for alternating currents above 1,000 V
IEEE Std 524-2003	IEEE guide to the installation of overhead transmission line conductors
IEEE Std 1048-2003	IEEE guide for protective grounding of power lines
IEEE Std. 1243-1997	IEEE guide for improving the lightning performance of transmission lines
IEEE Std C62.41.2-2002	IEEE guide for on-site acceptance tests of electrical equipment and system commissioning of 1,000 kV AC and above
National Electric Safety Code (NESC). 1997	C2-1997 National Electric Safety Code (NESC)
IEC 60099-4:2014	Surge arresters – Part 4: Metal–oxide surge arresters without gaps for AC systems
IEC 60099-5:2018	Surge arresters – Part 5: Selection and application recommendations
IEC 60099-6:2019	Surge arresters – Part 6: Surge arresters containing both series and parallel gapped structures – System voltage of 52 kV and less
IEC 60099-8:2017	Surge arresters – Part 8: Metal–oxide surge arresters with external series gap (EGLA) for overhead transmission and distribution lines of AC systems above 1 kV
IEC 60099-9:2014	Surge arresters – Part 9: Metal–oxide surge arresters without gaps for HVDC converter stations

protected by surge arresters, surge absorber and other protective devices. Different high-voltage tests and measurement techniques have been presented.

8.23 Standards or guidelines

Some useful standards and guidelines have been presented in Table 8.4 for further reading.

References

[1] K. O. Papailiou, Overhead Lines. CIGRE Green Books, Springer, Malters, Switzerland; 2017.

[2] Overhead Conductor Installation Guide Recommended Practices. 1st Edition, Installation guide by Electric Utility Engineering Section. General Cable Technologies Corporation, Kentucky, USA; 2014.

[3] P. N. Micropoulos and T. E. Tsovilis, 'Lightning attachment models and perfect shielding angle of transmission lines'. 44th Universities Power Engineering Conference, Glasgow, Scotland, 2009, pp. 57–70151.

[4] A. R. Hileman, Shielding of Transmission Lines Insulation Coordination of Power Systems. CRC Press, Taylor & Francis Group, New York, NY, USA; 1999, pp. 1–6.

[5] P. Sestasombut and A. Ngaopitakkul, 'Lightning protection of MEA's 24 kV distribution lines using overhead ground wires'. 2017 IEEE Innovative Smart Grid Technologies – Asia (ISGT-Asia), Auckland, 2017, pp. 1–5, doi: 10.1109/ISGT-Asia.2017.8378452.

Chapter 9
Insulation coordination

This chapter deals with coordination of insulation levels for overhead lines. As overhead lines operate at different voltage levels and connected with other parts of power system network, knowledge of power frequency nominal voltage levels has been presented. Then lightning of other transient over voltages has been presented. Concept of basic insulation levels has been provided mentioning its necessity. Then insulation coordination factors have been mentioned. Coordination with respect to insulation level of apparatus has been presented. Coordination with arresters has been described. Insulation coordination between overhead lines and connected substations has been discussed. At the end of the chapter, some useful standards have been listed for further study.

9.1 Introduction

Line conductors in overhead lines get separated from earth connection by insulators. Other metallic parts of support are earthed. Insulators installed in the overhead lines ensure smooth and fault proof operation for the transmission of electric power. Therefore, insulators are considered as the heart of overhead lines. Selection of type of insulators and their specifications with respect to electrical and mechanical parameters are important and must be carefully chosen to get desired performance from them. Insulation coordination refers to ensuring proper insulation levels as required by particular application with respect to other devices to be protected and reliable operation of protection units. Various related factors, voltage levels and insulation coordination have been discussed in the following sections.

9.2 Voltage factors insulation selection

Selection of insulation level in overhead lines depends on various mechanical and electrical factors. Among electrical factors, voltage is the most important parameter which draws main attention in the selection of insulation for particular application. Different voltage level indices consist of power frequency nominal voltage, power frequency peak voltage, lightning voltage peak, switching surge peaks, etc. Voltage levels important for insulation coordination are as follows:

- Voltage withstanding capacity of insulation
- Discharge voltage level of lightning arrester

9.3 Voltage signals in overhead lines

Overhead lines experience different types of voltage signals. They have been discussed as follows:

9.3.1 Power frequency operating voltage and power frequency over voltage

Power frequency nominal or operating voltage refers to the voltage level along with its frequency at which the overhead line connected in a power system network is expected to operate. Some commonly used power frequency nominal voltage levels are 400 V, 6.6 kV, 11 kV, 33 kV, 66 kV, 110 kV, 132 kV, 220 kV and 400 kV.

However, allowable voltage level is slightly higher than this nominal value which can occur at lightly loaded condition having poor regulation. Highest power frequency over voltage refers to the voltage level having frequency equal to the power frequency which exceeds the normal power frequency operating voltage and beyond which protection starts. This higher value is referred to as highest system voltage. For example, system with power frequency nominal voltage level of 400 V and 11 kV can have highest system voltage 440 V and 12 kV, respectively.

9.3.2 Power frequency voltage transients

Voltage transients showing a sudden rise of voltage having frequency equal to the range of power frequency is known as power frequency voltage transients.

9.3.3 High-frequency voltage transient

High-frequency voltage transient refers to voltage transients or a sudden rise of voltage peak(s) having frequency range higher than power frequency. High-frequency voltage transient may be of two types – multiple-cycle type transient and single-cycle type transient.

9.3.4 Direct lightning voltage

Lightning voltage is unidirectional type having low rise time and slower tail of decay.

9.3.5 Lightning restrike voltage

Lightning restrike voltage is unidirectional type with slower rise and decay time than that of direct lightning voltage.

9.4 Insulation coordination for a line insulator

Line insulators have different levels of voltage withstanding capability with respect to different types of voltage signals. In practice, withstanding capability of peak level of a particular type of voltage signal should be higher than the peak level of that particular type of voltage signal that can appear in the overhead lines. Different features of three types of suspension insulators have been provided in [1] with the following dimension as given as follows:

- IEEE Disk: 25.4 × 14.6 × 30.5 cm
- ANSI C292 Class 52-3: 25.4 × 14.6 × 29.21 cm
- Typical Fog Type: 25.4 × 14.6 × 43.18 cm

The number of insulators increases with the increase of voltage levels. For example, a commonly used number of insulators of type ANSI C292 Class 52-3 for different system voltages are 7–9 for 138 kV, 10–12 for 230 kV, 15–18 for 345 kV, 24–27 for 500 kV and 30–35 for 765 kV.

9.5 Basic impulse insulation level

Overhead lines and other power system equipment operate at different voltage levels. For different operating voltage levels, withstanding capabilities of insulations are different. For practice, the limit of insulation level has been set for different voltage levels of impulse waves of very short duration. This is known as insulation level. Standard basic impulse insulation level (BIL) for different class reference voltage level (kV) refers to minimum insulation level in terms of withstanding capability of crest voltage level of an impulse wave of short duration (1.2–5 μs). System or equipment insulation level must be capable of withstanding crest impulse voltage level above standard BIL. Some sample BIL with respect to reference voltage levels are presented in Table 9.1.

9.6 Insulation coordination with lightning arrester

Lightning arrester gives protection against peak levels of different types of high-voltage signals. To ensure reliable execution of lightning arresters, discharge voltage level in lightning arrester required for protection must be less than the voltage withstanding capacity of the line insulator as well as must be less than peak value of the high-voltage signals against which the arrester has been installed and that can arise in the overhead lines.

9.7 Substation considerations

High signals appear in the overhead lines can propagate to substation and vice versa. Therefore, proper insulation levels as well as protective measures are to be insured

Table 9.1 BIL with respect to reference voltage levels

Nominal power frequency voltage (kV)	33	66	132	220
Reference voltage levels (kV)	34.5	69	138	230
BIL (kV)	200	350	650	1,050

for both overhead lines and the substation connected with it. Sometimes, a lightning arrester of substation acts as back-up protection for itself installed in overhead lines or vice versa. In the same way, the insulation level of a line insulator may be less or higher than the insulation level of insulators used in substation. For overhead lines carried by wooden poles, line insulation level is normally higher than that of the substation connected with the line. For steel poles of towers, the insulation level of substation is higher than that of the overhead lines connected with that substation.

9.8 Common consideration for insulation coordination

Insulation coordination must be considered from the time of design. As overhead lines are connected with substations, generating stations, grid and consumers of different voltage levels, careful study of the connected network should be carried out and accordingly, voltage levels of each part of overhead distribution and transmission systems are to be decided. Some import considerations are mentioned as follows:

- Power frequency nominal voltage level
- Highest system voltage level
- Reference class kV
- Basic impulse insulation level (BIL)
- Insulation type
- String insulation level
- Application area of insulation
- Weather condition
- Lightning
- Other transients
- Protective devices

9.9 Contamination

Contamination levels in a specific area must be assessed before installation. Contamination materials found may be divided into two main groups:

- Conductive materials
- Inert materials

Contaminations that help in conduction are different types of metallic salts like sodium sulphates, sodium chloride, magnesium chloride, etc. Depending on the insulation type, shapes, etc. of insulating materials, these types of contaminations have their effects in flash over by creating different partial or full path of current. This type of contamination materials is soluble in water and changes dielectric property of the area in solution.

Contaminations that are found chemically inert are cement, clay, different oxides of silicone, etc. They do not chemically react; however, they can influence the insulation property. They form a layer that may hamper the natural process for washing the surface.

After installation, contamination materials should be collected at regular intervals for investigation and for further necessary steps, if needed. Specific leakage distance and stress withstanding capacity vary with contamination. In the presence of contamination, withstanding capacities are different for ceramic composite and porcelain insulators. Often washing is followed in silicone composite insulators [1]. Selection of water pressure is important for hot line washing. Water pressure applied may be low (less than 2,100 kPa), medium (2,100–2,800 kPa) and high (2,800–6,900 kPa). For ceramic and bonded polymer materials, washing by water of low-to-high pressure are not harmful. However, for unbonded polymer materials, water at low pressure is applied for washing. Sometimes, special silicone polymer, room-temperature-vulcanized liquid are used to improve performance in the presence of contamination [2–4].

9.10 Summary

This chapter has presented the major voltage parameters required for insulation coordination. Different types of high-voltage signals that may appear in overhead lines have been presented. Relative voltage levels have been discussed for insulation coordination between insulators, high-voltage signals and operating limits of lightning arresters.

9.11 Standards

Different useful IEEE and ANSI standards for insulation coordination in overhead lines have been presented in Table 9.2, and IEC standards have been presented in Table 9.3 for further study.

Table 9.2 Useful IEEE/ANSI standards for insulation coordination

Standard	Purpose
IEEE Std. 4-1995, Mar. 1995	IEEE standard techniques for high-voltage testing
IEEE Std. 987-2001, May 2002	IEEE guide for application of composite insulators
IEEE Std. 957-2005, 2005	IEEE guide for cleaning insulators
IEEE Std. 1523-2002, 2003	IEEE guide for the application, maintenance and evaluation of room temperature (RTV) silicone rubber coatings for outdoor ceramic insulators
ANSI Std. C29.13-2000, Jun. 2000	American National Standard for Insulators – Composite – Distribution Dead-end Type
ANSI Std. C29.7-1996, Jul. 1996	American National Standard for Wet-Process Porcelain Insulators – High-Voltage Line-Post Type
	American National Standard for Insulators Composite – Distribution Line Post Type

Table 9.3 Useful IEC standards for insulation coordination

Standard	Purpose
IEC 60168:1994	Tests on indoor and outdoor post insulators of ceramic materials or glass for systems with nominal voltages greater than 1,000 V
IEC 60168:1994	Tests on indoor and outdoor post insulators of ceramic materials or glass for systems with nominal voltages greater than 1,000 V
IEC 60168:1994	Tests on indoor and outdoor post insulators of ceramic materials or glass for systems with nominal voltages greater than 1,000 V
IEC 60273:1990	Characteristics of indoor and outdoor post insulators for systems with nominal voltages greater than 1,000 V
IEC 60305:1995	Insulators for overhead lines with a nominal voltage above 1,000 V – ceramic or glass insulator units for AC systems – characteristics of insulator units of the cap and pin type
IEC 60372:2020	Locking devices for ball and socket couplings of string insulator units – dimensions and tests
IEC 60383-1:1993	Insulators for overhead lines with a nominal voltage above 1,000 V – Part 1: Ceramic or glass insulator units for AC systems – definitions, test methods and acceptance criteria
IEC 60383-2:1993	Insulators for overhead lines with a nominal voltage above 1,000 V – Part 2: Insulator strings and insulator sets for AC systems – definitions, test methods and acceptance criteria
IEC 60433:1998	Insulators for overhead lines with a nominal voltage above 1,000 V – ceramic insulators for AC systems – characteristics of insulator units of the long rod type
IEC 60437:1997	Radio interference test on high-voltage insulators
IEC 60471:2020	Clevis and tongue couplings of string insulator units – dimensions
IEC 60507:2013	Artificial pollution tests on high-voltage ceramic and glass insulators to be used on AC systems
IEC TR 60575:1977	Thermal–mechanical performance test and mechanical performance test on string insulator units
IEC 60660:1999	Insulators – tests on indoor post insulators of organic material for systems with nominal voltages greater than 1,000 V up to but not including 300 kV
IEC 60720:1981	Characteristics of line post insulators
IEC TR 60797:1984	Residual strength of string insulator units of glass or ceramic materials for overhead lines after mechanical damage of the dielectric
IEC TS 60815-1:2008	Selection and dimensioning of high-voltage insulators intended for use in polluted conditions – Part 1: Definitions, information and general principles
IEC TS 60815-2:2008	Selection and dimensioning of high-voltage insulators intended for use in polluted conditions – Part 2: Ceramic and glass insulators for AC systems
IEC TS 60815-3:2008	Selection and dimensioning of high-voltage insulators intended for use in polluted conditions – Part 3: Polymer insulators for AC systems

(Continues)

Table 9.3 (Continued)

Standard	Purpose
IEC TS 60815-4:2016	Selection and dimensioning of high-voltage insulators intended for use in polluted conditions – Part 4: Insulators for DC systems
IEC 61109:2008	Insulators for overhead lines – composite suspension and tension insulators for AC systems with a nominal voltage greater than 1,000 V – definitions, test methods and acceptance criteria
IEC 61211:2004	Insulators of ceramic material or glass for overhead lines with a nominal voltage greater than 1,000 V – impulse puncture testing in air
IEC TS 61245:2015	Artificial pollution tests on high-voltage ceramic and glass insulators to be used on DC systems
IEC 61325:1995	Insulators for overhead lines with a nominal voltage above 1,000 V – ceramic or glass insulator units for DC systems – definitions, test methods and acceptance criteria
IEC 61462:2007	Composite hollow insulators – pressurized and unpressurized insulators for use in electrical equipment with rated voltage greater than 1,000 V – definitions, test methods, acceptance criteria and design recommendations
IEC 61466-1:2016	Composite string insulator units for overhead lines with a nominal voltage greater than 1,000 V – Part 1: Standard strength and end fittings
IEC 61466-2:1998	Composite string insulator units for overhead lines with a nominal voltage greater than 1,000 V – Part 2: Dimensional and electrical characteristics
IEC 61466-2:1998	Composite string insulator units for overhead lines with a nominal voltage greater than 1,000 V – Part 2: Dimensional and electrical characteristics
IEC 61467:2008	Insulators for overhead lines – insulator strings and sets for lines with a nominal voltage greater than 1,000 V – AC power arc tests
IEC 61952:2008	Insulators for overhead lines – composite line post insulators for AC systems with a nominal voltage greater than 1,000 V – definitions, test methods and acceptance criteria
IEC 61952-1:2019	Insulators for overhead lines – composite line post insulators for AC systems with a nominal voltage greater than 1,000 V – Part 1: Definitions, end fittings and designations
IEC TS 62073:2016	Guidance on the measurement of hydrophobicity of insulator surfaces
IEC 62155:2003	Hollow pressurized and unpressurized ceramic and glass insulators for use in electrical equipment with rated voltages greater than 1,000 V
IEC 62217:2012	Polymeric HV insulators for indoor and outdoor use – general definitions, test methods and acceptance criteria
IEC 62231:2006	Composite station post insulators for substations with AC voltages greater than 1,000 V up to 245 kV – definitions, test methods and acceptance criteria

(Continues)

Table 9.3 (*Continued*)

Standard	Purpose
IEC 62231-1:2015	Composite station post insulators for substations with AC voltages greater than 1,000 V up to 245 kV – Part 1: Dimensional, mechanical and electrical characteristics
IEC TS 62371:2008	Characteristics of hollow pressurized and unpressurized ceramic and glass insulators for use in electrical equipment with rated voltages greater than 1,000 V
IEC TR 62662:2010	Guidance for production, testing and diagnostics of polymer insulators with respect to brittle fracture of core materials
IEC TR 62730:2012	HV polymeric insulators for indoor and outdoor use tracking and erosion testing by wheel test and 5,000 h test
IEC 62772:2016	Composite hollow core station post insulators for substations with AC voltage greater than 1,000 V and DC voltage greater than 1,500 V – definitions, test methods and acceptance criteria
IEC TS 62896:2015	Hybrid insulators for AC and DC for high-voltage applications – definitions, test methods and acceptance criteria

References

[1] A. C. Baker, M. Farzaneh, R. S. Gorur, *et al.*, Working Group on Application and Inspection of Insulators, 'Insulator selection for AC overhead lines with respect to contamination.' In IEEE Transactions on Power Delivery, vol. 24, no. 3, pp. 1633–1641, 2009.

[2] "IEEE Guide for the Application, Maintenance, and Evaluation of Room Temperature Vulcanizing (RTV) Silicone Rubber Coatings for Outdoor Ceramic Insulators." In IEEE Std 1523-2002, pp.1–28, 2003. doi: 10.1109/IEEESTD.2003.94238.

[3] J. J. LaForest, Transmission Line Reference Book—345 kV and Above (2nd Edition). Elect. Power Res. Inst., Palo Alto, CA, USA; 1982.

[4] J. N. Edgar, J. Kuffel, and J. D. Mintz, Assessment of the Contamination Performance of Transmission Class Composite Insulators Using the Clean Fog Test Procedure CEA Project 280T621. Can. Elect. Assoc., Montreal, Canada; 1993.

Chapter 10
Route selection, commissioning, operation and maintenance

This chapter deals with route selection, commissioning, operation and maintenance of overhead power transmission lines. It starts with route selection followed by general considerations of understanding of purpose, resource and objective, covered area, geographical and geological diversity, political map, climatic statistics, cost study, etc. Different aspects of establishing new lines, alternate lines, lines conversion have been discussed. Linking with underground lines has been discussed. Different aspects of planning and construction like survey, planning, design, foundation, installation, erection, commissioning, responsibility and issues, check-up, test, energization, etc. have been discussed. Supervision and quality assurance in commissioning and commencement of operation have been discussed thoroughly. Operation and maintenance, post-commissioning planning and management, operation management, maintenance management, asset management, risk management have been discussed followed by uprating, upgrading and extension of line.

10.1 Introduction

Electric overhead lines are built and expected to have long life. It shares major component of cost of overall power system. It involves with various social–economic–geopolitical issues. Therefore, actual process of building power system network with overhead lines starts much before its inception. Various major steps for overhead electric lines since starting have been shown with a block diagram in Figure 10.1.

Project work starts with suitable route selection. Then, planning and construction starts. After construction work, commissioning is done. Operation and maintenance continue throughout the rest of the life – sometimes in cascade mode and sometimes simultaneously.

10.2 Route selection

Route selection is the first important job for building of overhead line. It starts much before planning stage. In finalization process, various factors are to be scrutinized. Inference with different possible routes is placed in the form of report to the decision-making authority which may be either public body or private

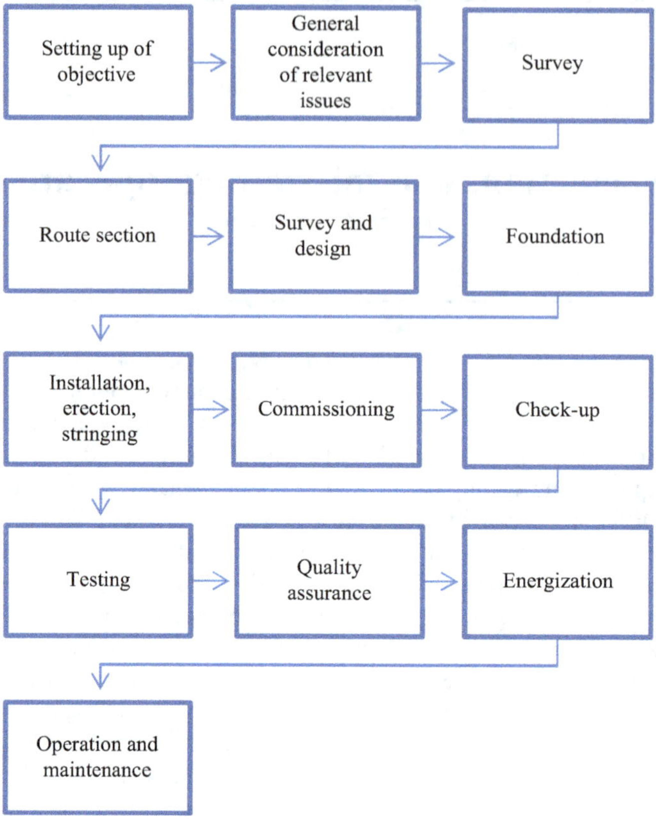

Figure 10.1 Schematic diagram for various steps for project of making overhead lines

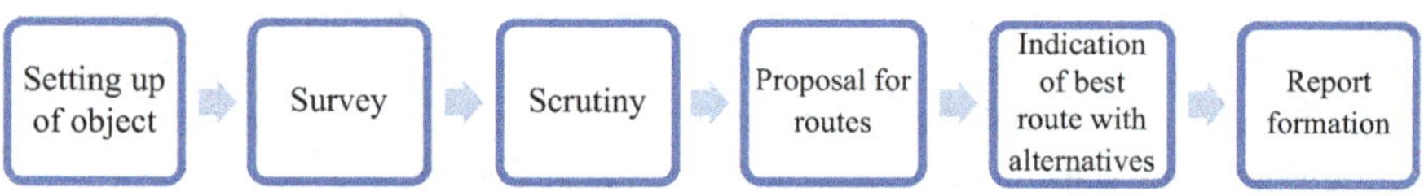

Figure 10.2 Schematic diagram for various steps for rout selection overhead lines

organization for taking final decision and approval. Steps of route selection have been shown with a schematic block diagram in Figure 10.2.

- *Setting up of objective or purpose:* First, main objective or purpose should be setting up. In most of the cases, this objective is given by the authority and there is little scope to change it.
- *Survey and scrutiny of various factors:* Survey and study of various factors, which have direct or indirect influence on route, are to be carried out. All outcomes of surveys and studies are consolidated and scrutinized.

- *Proposal of different routes:* After scrutiny, different possible routes are proposed.
- *Indication of best route and next few alternatives:* Many routes for installation of transmission line may be possible. It is then needed to find out the best fit route, and along with best solution, few alternatives are to be indicated in order.
- *Report formation:* Final report is to be prepared along with solutions with supporting causes and difficulties that may arise. Now the report is to be placed to the authority.

10.2.1 General considerations

Different factors are considered in the perspective of objective or purpose. Normally, it is given by the authority and there is little scope to change it. Survey and study are made on various factors that may have direct or indirect influence on route. Historical statistics are also to be studied with this respect. All outcomes are analysed to find out possible routes for the specific purpose. Following major factors are considered in route section:

1. Understanding of purpose
2. Resource and object ends
3. Covered area
4. Geographical and geological diversity
5. Political map
6. Average climatic statistics
7. Probabilistic extreme climatic condition
8. Cost study

10.2.1.1 Understanding of purpose

Clear understanding must be there on the purpose of route selection. The people involved with route selection may not have the authority to change the purpose, and they have to go ahead by the target. However, at the end, they can propose for slight change or resetting of purpose, if required and found essential. Clear understanding about purpose gives clear picture of beneficiaries and relation with all stake holders. The purpose may include trade relation between public and private companies or between two government authorities of different states and provinces or between two or more countries.

10.2.1.2 Resource and object ends

Overhead lines have minimum two ends or may have multiple end points. This depends on the resource of power and objective of power transmission and distribution. For transmission system, there will generally be two or more substations located in different regions. In the case of distribution line, there may be two or more substations of the network at the customer ends. This information will also make it clear about the voltage level of overhead lines to be considered. Accordingly, other parameters will have to be judged like support height, clearance, sag, safety.

10.2.1.3 Covered area

When ends of the overhead lines are identified, covered area may be located. This may include the same state or province where the same electricity guidelines work. This may also include different states or provinces where different rules are governed by the regulatory bodies. This may include two or more different countries which may be governed by completely different standards and guidelines. Attempt is normally to be taken to cover the areas which will raise minimum number of conflicts and difficulties with respect to rules, regulations and standard aspects.

10.2.1.4 Geographical and geological diversity

When the cover areas are tentatively identified, existing geographical and geological diversities are found. These will give idea on practicability of overhead lines of particular voltage levels in the area. This may include one or many of the followings:

- Plain land
- Hilly area
- Desert
- Areas with lots of rivers of different nature
- Deccan areas
- Forest
- Deep forest
- Volcanic areas
- Seas or sea banks

Geological information will help in finding out possible foundation types and their feasibility. For example, foundation type varies with the variation of soil type. Other than electrical parameters, structure depends on soil strength, slope of the land, etc. Clearance of the conductor from the ground level depends on the sag. Natures of sag in plain land and slope or hilly areas are different. There may be a need of cutting or trimming of trees in the right of way of the overhead lines especially in the forest areas. It may be noted that overhead line may pass through areas of different geographical and geological patterns. Therefore, foundation, support structure, sag, etc. may vary in different spans of an overhead line.

10.2.1.5 Political map

Political map of the covered area must be studied. This relates regulations and standards applicable in particular regions. This study may also drive the direction for solving some conflicts that may arise during planning and construction.

10.2.1.6 Average climatic statistics

Long-time average climatic statistics are to be collected; this will help in deciding feasibility of particular route with respect to technical alternatives. For example, in some route, ice is not common or not found in the history of, say, 100 years; whereas the area may experience severe rain or it may be dry throughout the year. As technical specifications are very much dependent on the climatic conditions, in the route selection process, the path which involves comparatively known climatic condition and minimum diversity is chosen.

10.2.1.7 Probabilistic extreme climatic condition

Some climatic conditions show extreme behaviour. Therefore, in route section, probabilistic analysis should be carried out. Some extreme conditions, which have not seen in last 30 years, may appear in next 30 years. Some pocket geographical areas are found, which shows extreme climatic behaviour. If possible, these areas may be avoided.

10.2.1.8 Cost study

Cost involved in setting up the route varies. In some cases, route may cover wide rivers of weak soil which requires special attention for foundation, or deep forest or agricultural lands which need a separate type of attentions. Owners of the land may be different – public and private; and different rules may be there to utilize the respective land. In some cases, land owners' compensation may arise. All these involve with the cost in setting up route in reality. Therefore, assessment on cost involved in setting up route should be carried out.

10.2.2 Guidelines

With the progress of society, electric power demand has increased to a great extent. In fact, it has come to the index of growth of civilization. Thus with the increase of demand advancement of technology has occurred. Various types of multi-circuit overhead lines of high voltages along with HVDC have been introduced. Based on the past experience, a new task is expected to both of the followings:

1. Standards
2. Guidelines of regulatory bodies

10.2.2.1 Standards

Standards on different aspects of overhead lines have been introduced by international bodies. Some have been presented in Table 10.1 for further study. Standards are updated after a certain period of years when requirement comes. Therefore, while referring to particle standards, its updates should also be followed.

10.2.2.2 Guidelines of regulatory bodies

Regulatory bodies are mainly formed by respective state or province or country. Government set regulatory bodies on power sector. These bodies generally follow suitable internal standards and accordingly set guidelines for the purpose. In starting a project on overhead lines, at each stage, guidelines are to be noted and followed. There may exist some cases with no guidelines. In those cases, it is to be checked whether there is any standard practice on issue and what are their experiences. Overhead line projects may involve with the followings:

- Completely new line
- Setting up alternate path of electric power
- Line conversion

Table 10.1 Useful standards on overhead lines

Body	Standard	Purpose
ANSI/IEEE	ANSI/IEEE Std 516-1987	Guide for maintenance methods on energized power lines
IEEE	IEEE Std 524-2016	Guide for the installation of overhead transmission line conductors
IEEE	IEEE Std 664-1980	Guide on the measurement for the performance of Aeolian vibration dampers for single conductors
IEEE	IEEE Std 738-1993	Standard for calculating the current-temperature of bare overhead conductors
IEEE	IEEE 751-1991	Trial-use design guide for wood transmission structures
IEEE	IEEE Std 977-2010	IEEE guide to installation of foundations for transmission line structures
ANSI/IEEE	ANSI/IEEE Std 987-1985	Guide for application of composite insulators
IEEE	IEEE Std 998-1996	Guide for direct lightning stroke shielding of substations
IEEE	IEEE Std 1025-1993	Guide to the assembly and erection of concrete pole structures
IEEE	IEEE Std 1410-1997	Guide for improving the lightning performance of electric power overhead distribution lines
IEEE	IEEE Std 1628-2009	Recommended practice for maintenance of DC overhead contact systems for transit systems
IEEE	IEEE Std 1724-2011	Guide for the preparation of a transmission line design criteria document
IEEE	IEEE Std 1829-2017	Guide for conducting corona tests on hardware for overhead transmission lines and substations
IEEE	IEEE Std 1863-2019	Guide for overhead AC transmission line design
IEEE	IEEE Std 1893-2015	Guide for the measurement of DC transmission line and earth electrode line parameters
IEEE	IEEE Std C135.63-1998	Standard for shoulder live line extension links for overhead line construction
IEEE	IEEE Std C2-2007	National Electrical Safety Code 2007 Edition
ANSI	ANSI C135.31-1988	American National Standard for Zinc-Coated Ferrous Single And Double Upset Spool Insulator Bolts for Overhead Line Construction
ANSI	ANSI C135.22-1988	American National Standard for Zinc-Coated Ferrous Pole-Top Insulator Pins with Lead Threads for Overhead Line Construction

New lines

In setting up new lines, engineers generally have more flexibility than others. They sometimes have the flexibility of setting voltage level and then can find out the route and proceed as per the standards and guidelines available to them. Sometimes in the project of setting one or multiple number of lines, planner can include in the proposal of setting up intermediate substation in addition to terminal substation.

Target terminal locations are normally fixed and given by the either government authority, or public and private companies. However, the path connecting two or more areas may be changed or may have multiple options in setting up new lines.

Alternate lines
In setting up alternate overhead lines, first and foremost, we have to make study and survey on the existing line whose alternate is being made. Engineers will have to find out what are the possible best practices that can be incorporated in setting the alternate lines. Sometimes, alternate overhead lines may be possible on the same support structure by adding or changing the cross arm design feature. Sometimes, new supports are needed. In some cases, the same route may be followed, in some other cases better and more cost optimum route may be found.

Line conversion
Line conversion refers to setting up either a new route or new lines or both, establishing new overhead lines and to obsolete old lines. This line conversion is different from voltage conversion that is done in substation or conversion from AC to DC or vice versa. It may be considered as the replacement of old lines which is either damaged in major part or obsolete due to ageing effect and is not getting 'fit certificate' with respect to present day need of power system network at large. Line conversion is therefore bounded by some boundary conditions that are valid for past line. Engineers change it to offer more feature, flexibility, power transmitting capacity and of course enhanced stability limit. The rejected line may be, say 40 years old; now then the line is being replaced. The population, environment and infrastructures in the covered area normally undergo big changes which must be considered in handling line conversion project.

10.2.3 Linking with underground cables

Linking with underground cables is sometimes required while crossing roads, rails, rivers, area of higher population density. In practice, these are seen in distribution lines which run in the areas of city or suburban areas. However, such link line is not very high with respect to the length of overhead line. Modern distribution to the consumers in cities is made through underground lines. However, they may be connected with overhead lines. This meeting may or may not occur at substation.

10.3 Planning and construction

After selection and finalization of route of overhead lines of any category (new, alternate or conversion type), planning starts followed by construction. Planning and construction of overhead lines involves with following stages:

1. Survey
2. Planning
3. Design

4. Foundation
5. Installation
6. Erection

10.3.1 Survey

Survey for planning works is different from that is done during route selection. Survey is to be carried out considering the route finalized. Different technical aspects, with geographical and geological information along this route, will still be useful in this survey. Voltage levels and technical constrains are to be studied along with their flexibility to change, if required.

10.3.2 Planning

In planning, all aspects are to be judged in terms of feasibility. Optimum cost-effective planning is always welcome. Planning includes

- Route selection
- Budget
- Time for execution
- Step-wise time allocation for foundation, installation, erection
- Technical drawing (includes civil, mechanical and electrical)
- Guidelines for commissioning

10.3.3 Design

Sometimes, design is carried out after planning or some times, and it is carried out as a part of planning. Design part includes all civil, mechanical and electrical drawings of the projects, specifications of all devices, members and sub-members of the overhead lines. It involves with design of support, line conductor, earthing, earth wire and protection against lightning. Other protection may be included from starting or may be included gradually. Each step of design should have diagrams with proper scale and all technical dimensions.

10.3.4 Foundation

Foundation works is planned and designed based on the soil. Various types of soil are available in different places. Some are highly suitable for the foundation of towers, some are found not suitable for foundation of tower. Soil in river side and hilly areas are different from soil of plain lands. In the area covered by one overhead line, there may be the same type or different types of soils. Therefore, the foundation of support structure may be the same or may be different from each other; before starting planning and foundation works, prior knowledge about the soil is necessary.

Soils are broadly classified into the following three categories:

1. Undisturbed soil
2. Rock soil
3. Artificial soil

Based on height of the support, various loads on support, soil type and soil strength, foundation work of electrical supports is carried out in two ways:

1. Compact foundation
2. Separate foundation

There are several types of compact and separate foundations. A particular type of foundation is selected based on various factors, including mechanical loads expected to withstand by the support and soil nature. Foundation work includes foundation of support as well as of guy. In addition, earthing to soil is also to be considered and done, simultaneously after foundation.

10.3.5 Installation

Installation work is carried out by a separate group of people. Guidelines and standards as prescribed by regulatory bodies must be followed in installation [1,2]. After foundation is over, installation of support or tower starts. Typical 3–10 days are required to complete a lattice tower. Work starts from base level coming from the end of the foundation. It starts with installation of legs, waist, tower body, cross arm and top, including various horizontal and diagonal arms. In this process, number, quality and proper dimension matching of all nuts and bolts are very important and should thoroughly be checked.

10.3.6 Erection

Conductors are connected with support in erection. This process involves with proper connection of insulators, fittings, stringing and position of line conductors and sky wire. Overhead conductors as specified by the designers are brought and used. Conductors of the same material or composite conductors are used. Both line conductors and sky wire are positioned. This process is guided by limits of tension, specified sag, clearance and distance between conductors. If roads or other networks exist under the path of overhead lines, all cares should be taken during the process without hampering normal works. All danger notices and safety measures must be taken for others as well as for people who are engaged in the process.

10.4 Commissioning

Commissioning refers to a supervision of all types of working in the process initiating from the time of approval to energization of the line. Commissioning deals with non-technical issues as well as issues related to civil, mechanical and electrical issues involved with overhead lines. Care is taken so that no small part remains unnoticed. Testing of foundation is carried out. Starting from all nuts and bolts, all members of tower are checked. Guys are checked. Strings, conductor tension, sag and clearances are checked. It must include checking of earthings in the line, sky wire for lines of high-voltage lines and neutral lines for low-voltage distribution lines. The presence and quality of guard wires are also considered in commissioning. In fact, these are just few to mention; care is taken so that no part remains unnoticed in commissioning. Thus, aim of commissioning is quality assurance in all respect of relevant standards, safety measures and agreements with the line owner and other stake holders before starting full-fledged operation.

10.4.1 Responsibility and issues

Responsibility of carrying out commissioning job primarily goes to the line owner itself. But the line owner may not be an expert in doing this job or may not have optimized and feasible working group in this area. Therefore, it is often observed or advised to handover the responsibility of commissioning to the members of an independent third party who are expert and have experience in carrying out commissioning in the recent past. In the commissioning process, all issues involved with commissioning are listed. Issued are included as per the guidelines, standards and design criteria of the overhead project work. As said, generally commissioning is carried out by a separate group of independent people who are authorized for quality assurance. It is also related with business conditions, payment to vendors and giving a clearance to the respective stake holder to start operation.

A line owner should make an agreement with the commissioning company, including all aspects of desired outcomes. For this purpose, they can depute professional law firms. A commission company must report to the line owner at intermediate stages and after completion of jobs. One major reason in business aspect is that often payment or part payment to the suppliers, manufacturers and different working companies depends on the report of commissioning. All reports of commissioning should be filed and recorded because they may be referred in future for many interests. Often many banks or companies invest in the project. Therefore, all of them may rely on the same commissioning group, or an individual investor may form own commissioning of customized purposes.

10.4.2 Check-up

All issues as listed in agreement followed normal practices and guided by regulations are to be checked one by one. For this purpose, a checklist must be prepared, and it must be submitted to the concerned authority after finishing jobs. It starts from supply of the materials, manufacturing and foundation. It then includes all jobs of erection, insulator installation, stringing, and protection and safety measures. All safety measures are checked thoroughly and with the protective measures taken against natural calamities, lightning and other electrical disturbance. Checking commonly includes the following:

- Whether the approval process includes all related rules, regulation and public laws in the area. This may be done with law experts.
- Land issues: This issue must be paid attention before starting the work based on the laws of the provinces and public interest.

- Public issues connected with this like crossing roads, rails, communication lines, rivers.
- Possible temporary access to other lands that needs to be used during the process and issues with this, including compensation of the damages occurred due to this like loss of agricultural property, forests and assets.
- Whether suppliers are supplying materials and subcomponents in time.
- Number and quality of supply materials. Some of these may not been done after installation in a feasible way.
- Manufactured materials or components in terms of number and quality.
- Foundation type, specifications and quality.
- Earthing quality: Soil resistance, earthing resistance and earthing arrangements.
- Erection finishing: This includes nuts and bolts, members of tower, clamps, etc. in terms of numbers and quality.
- Level of insulation at relevant points: insulation levels with respect to line voltages, lightning and other over voltages that may appear in the overhead lines.
- Conductor quality: Matching of specification and sample tests, company supplied papers.
- Conductor damage: It includes damage of stranding, cores and surface areas mainly in handling and stringing process.
- Fibre cables, if involved: Mainly involved with sky wire like optical ground wire for communication purpose and its proper functioning.
- Guy and guard wires: Foundation of guy and clamping of guy and guard wires and ground wherever applicable.
- Sky wire: Its positioning, tension, sag and clearance from ground and line conductors.
- Corona protection: Proper installation of corona ring and other measures.
- Safety measures with respect to all parts of support, auxiliary parts, surrounding areas of support, areas located under the line, public areas like roads, rails, communication tower or lines.
- Safety awareness like danger notice.
- Protective measures against lightning [3,4] and other surges, etc. Shielding angle [5] is checked.
- Whether unused parts of wastage materials have been removed from the area of installation and its proper reporting. No such part should be burnt or left in the place.

10.4.3 Test

Where visual inspection or measurements are not sufficient, tests are carried out. For a particular technical aspect, sample tests or random tests may be followed. It should be noted that there is no lower limit or upper limit of number of tests. As many as possible tests may be carried out to establish confidence on safety, quality and all other relevant issues. Some common tests followed are as follows:

- *Foundation test:* Sample test for strength and stability assessment. Foundation tests and inspection varies based on the type of foundations (like compact or isolated) soils and locations (like plain land, river, hilly areas).
- *Earthing:* It refers to measurement of earthing resistance, continuity of earthing, whether earthing has included all metallic connection other than the line conductor connection, auxiliary parts like fencing, guards.
- *Conductor resistance:* It is done by measuring conductor average resistance at certain temperature with small DC voltage of 12/24 V. Then, conversation of resistance at 20°C is made and the value is compared with the set value.
- *Power loss:* It refers to determination of power loss with the help of resistance value.
- *Conductor motion:* It refers to vibration test that includes aerial and sub-span motion.
- *Insulator test:* It refers to insulation levels in different conditions. It is to be done as guided by the insulation coordination of the overhead lines and interconnected substations.
- *Energization:* Testing ends with successful energization. Sample energization is made after taking permission of line owners, all clearance and safety measures.

10.4.4 Energization

Energization in the line is done after thorough check-up, tests and inspection of protection and safety measures. Prior clearance is required to do this. Energization is also referred to as charging of overhead lines. The process involves with coordination of sending end and receiving end substations connected by the overhead lines. Therefore, it may involve different companies or authorities. If the overhead line covers two or more different countries, it involves with clearance from each of the countries regulatory body; especially for transmission line. For distribution line, charging this binding is comparatively less. By energization, power flow and functioning of all parts (mainly electrical) are judged. Along with this conductor, motion is monitored. Functioning of insulation, earthing system and all protective measured are checked. If any failure or shortcoming is observed, the process is stopped at the earliest and again inspection, testing and steps for solving the issues are made. After this, again the line is energized or charged to look back all checklists.

10.4.5 Supervision, quality assurance in commissioning and commencement of operation

Supervision is part of commissioning involved at each stage of the project of overhead lines. This work is done with checklist made as per guidelines, standards and practice. Each point or issue must be verified and consolidated as report form. All these are tallied to ensure quality involved in outcomes expected to achieve. Normally, it is submitted to a line owner or his/her authorized agency. Quality audit is done and if quality standard reaches up to the mark, then approval request for commencement of operation is made to the competent regulatory body in the state or country.

10.5 Operation and maintenance

After successful commissioning, overhead line is made ready for operation. Some maintenance jobs are carried out simultaneously, some maintenance jobs are performed, after isolating the overhead line.

10.5.1 Operation

Operation part is mainly done by electrical engineers whereas maintenance jobs involve electrical, mechanical and sometimes civil engineers. In operation, overhead lines are expected to show desired characteristics of load flow, including voltage profile and electrical stability. Along with this, the overhead line should show a high level of reliability against various types of electrical faults and provide all types of safety measures. During energization and initial stage of operation, insulation, temperature and vibration of conductors must be monitored. They may be included as part of normal investigation.

10.5.1.1 Insulation

Effect of sudden energization, normal loading and other switching operations on overhead line insulator are closely monitored at the stage in first energization and initial stage of operation. If any failure or abnormalities are noticed, attempt is taken for the replacement of a particular insulator unit.

10.5.1.2 Temperature of conductor

Temperature of conductors may be monitored in a direct or indirect way. By an indirect way, conductor current is monitored and temperature is calculated by power loss and heat dissipation. In direct method, temperature sensors are applied, sensors are used to measure both conductor temperature and atmospheric temperature. Time–temperature data are monitored for long duration and stored for analysis.

10.5.1.3 Motion of conductor

Initially, aeolian and span motions of a conductor are monitored. This monitoring is also a long-term process and wind speed dependent. Effects of other factors on conductor vibrations are also noted. Galloping found in overhead lines is comparatively rare. For galloping, one may have to wait for long. Therefore, both vibration parameters and wind speed and directions are to be monitored with time. Time-wind speed-vibration data are monitored and stored for analysis.

10.5.1.4 Operation of safety and protective measures

Smooth and safe operation of lines is impossible without proper functioning of safety equipment and protective devices installed in different parts of the overhead lines. Overhead lines are exposed to open air and face frequent lightning and other over voltages. Line should be well protected from lightning [3,4] and other surges. Therefore, operation monitoring of safety measures and protective units are made in parallel and in synchronism with transmission of power through the line. Minor or major failure with this regards leads to urgent maintenance.

10.5.2 Maintenance

As said, maintenance jobs may be electrical, mechanical or civil. Even without any problems, routine check and servicing of some parts are advisable. However, first few years, mechanical or civil related problems are found rare. Normally, in modern business relation, the same company is given the responsibility of carrying maintenance works involved in overhead lines. In tender and bidding process, condition of providing maintenance, for coming years of say 50–100 years, is included.

Maintenance jobs may broadly be divided into the following three categories:

1. Investigation
2. Service
3. Repair

10.5.2.1 Investigation

Investigation of overhead lines includes the following for each part of the line:

- Soil
- Foundation
- Support structure
- Insulation
- Guy
- Sky wire
- Earthing
- Lightning arrester
- Conductor
- Environmental effect
- Inference with other networks
- Interaction with surroundings

Investigations are planned on regular basis. Frequency and timings of investigation are scheduled, and different teams are formed for the inspection of different categories. At the end, all findings are consolidated with time and location and submitted in the form of report. At this, recommendation may be done but decision on corrective measures is not normally taken; they are placed to the competent authority for this. Sometimes investigation involves with long-time monitoring based on the behaviour of the overhead lines considering geodiversities and climatic variations. One particular overhead line may go through areas of different geographic and climatic natures (like plain land, sea or sea side, hilly areas, desert, very hot, extremely cold). Therefore, investigation issues of monitoring parameters may vary based on the locations of investigation of overhead lines. Modern investigation may also include an interference of other networks and interaction of surrounding objects with the line and their effects; all these should not be ignored.

10.5.2.2 Service

Service mainly involves mechanical jobs. Foundations, structures and all fittings of all parts of overhead lines are considered or checked whether and where service is required primarily based on investigation report. In most of the cases, service deals with following issues:

- Missing of nuts and bolts
- Loose fittings
- Burnishing, galvanization, colouring

10.5.2.3 Repair

Repair jobs in overhead lines include jointing, threading, fittings and replacement of different fittings or sub-members.

Service and repair works may be carried out under two conditions, viz.

1. Dead line
2. Live line

Dead line

Before carrying service or repair work line is completely de-energized. All parts are brought to zero potential. For additional safety measures, line conductors and substation terminals connecting particular section are earthed.

Live line

In live line work, the following attempts are taken only after taking all safety measures for the people at job:

- Raise the potential slightly above ground potential
- Use of insulated gloves
- Use of insulated sticks
- Use of insulted platform
- Use of a drone or a helicopter with proper insulation

In terms of management, there may be a problem of lack of skilled persons to work in live line. Moreover, it is often observed that workers do not prefer to work in live line. On the other hand, today's ever increasing power demand in deregulated market demands for uninterrupted power distribution. Therefore, if skilled man power willing to work in live lines is not found, then automated and mobile working station with robotics application having 6 degrees of freedom for movement and camera with three-dimensional views may be utilized for live line work.

10.6 Post-commissioning planning and management

Post-commission management roles for whole life span of overhead lines. It includes the followings:

- Operation management
- Maintenance management

- Asset management
- Risk management
- Uprating
- Upgrading
- Future planning

10.6.1 Operation management

Operation management often overlaps with substation management connected with the lines. It monitors and controls/distributes power flow, reactive powers, switching operations. Protective measures are also monitored and managed by operation management and dynamic management [6]. Utility of the line largely depends on operational management of a line. In operational management, monitoring and control plays important role. Overhead lines are characterized by ampacity, tensile strength, sag limits, etc. Therefore, along with the monitoring of electrical loads from substation ends, monitoring of the following is necessary:

- Temperature
- Tensile strength
- Sag
- Vibrations
- Effect of ice and wind
- Effect of thunder storm

10.6.1.1 Temperature

Temperature of an overhead line conductor is monitored with respect to surrounding temperature. Conductors' temperature may be measured by direct and indirect methods. In the direct method, temperature is monitored by sensors' application. In indirect approach, temperature is calculated from loss calculation, heat generation and heat dissipation. Effects of wind flow and surrounding temperature are also monitored as they have influence on conductor temperature.

10.6.1.2 Tensile strength

It is an important parameter needs to be monitored. Effects of wind flow and ice are considered in the determination of tensile strength. Upon tensile strength, mechanical loads on the support legs are decided.

10.6.1.3 Sag

Initially sag is calculated during the change of construction during operation. Also, it is influenced by ice and wind flow. Upon sag, clearance of the conductor from ground level is decided. Minimum clearance must be maintained as per guidelines of the regulatory body which is different for different voltage levels of the operating overhead lines. Overhead line may undergo different types of geographical areas and nature, and maximum magnitude of sag may vary.

10.6.1.4 Vibration

Aeoline and sub-span vibration are monitored. Along with this, performances of dampers are also checked. Galloping is observed from long-duration monitoring to see whether any overlapping between conductors is occurring in the line.

10.6.1.5 Ice and wind

Effects of ice and wind are observed throughout the year. In some regions, ice does not occur at all, in some areas both ice and wind effects are very prominent. Sag, mechanical strength, temperature, etc. are monitored along with fog, ice and wind speed and database is developed for immediate and future action.

10.6.1.6 Thunderstorm

Thunderstorm is special type of atmospheric phenomenon which does not occur every day. Therefore, monitoring the effect of thunderstorm is not possible suddenly. Throughout the year, weather is monitored and the number of lightning is counted. Performances of lightning arrester, sky wire, etc. are monitored.

10.6.2 Maintenance management

Maintenance management monitors and performs the jobs of investigation, service and repair keeping coordination with operation management. It has a routing unit and an emergency unit. In fact, reliability, health and longevity of overhead lines largely depend on this management. The owner of the lines prefers to depute a third party experienced in this particular management. As in modern age, all standard practices follow insurance of assets, this management team works in correlation with insurance company and keeps all records and future conflicts for further analysis.

10.6.3 Asset management

Asset management deals with the expenditure involved with planning, construction and operation. It includes cost involved in the following but not limited to

- Survey
- Planning
- Approval
- Design
- Land
- Foundation
- Materials manufacturing
- Support assets
- Conductors, insulators, protective and safety measures
- Construction, erection, commissioning
- Operation management
- Share prices
- Bank loan
- Present value

- Yearly expenditure
- Power trades shared by the line

10.6.4 Risk management

Risk management deals with the failure occurred in the line. It may be due to internal cause or external cause. It is assessed mainly by two following parameters:

- Number of failure risk per km
- Number of failure risk per year

Risk [7] depends on the failure numbers, category and their probability of occurrence. Initially, statistical information of similar type lines is helpful. Gradually, own data bank is made on the basis of failures occurred in own line. Probabilistic predication is made by the data, and precautionary measures are recommended. Risk management is correlated with all other managements like operation, maintenance, uprating and upgrading. Sometimes, risk management results in a proposal of upgrading and extension to reduce the risk probability of particular overhead line under consideration. Risk management is also linked with financial loss assessment involved with the line.

Right-of-way refers to the area above which overhead lines are positioned. Monitoring of right-of-way is usual practice under risk management. Vegetation index and snow index are two indices used to indicate vegetation level on the ground and amount of snow on the surrounding lines. These indices are monitored. To reduce the risk in lines, trees on the right-of-way are cut in normal practice.

Safe clearance in all respect should be ensured in risk management for the people passing through 'right-of-way', people working in construction or maintenance, overhead lines, interrupted power supply and surrounding properties. Therefore, probable risks are noted, monitored, and measures are to be taken to eliminate risk or reduce risk as much as possible to ensure safety as per guidelines.

10.6.5 Uprating

Uprating of overhead lines refers to resetting of line specification for defining operation range maintaining or ensuring safety, reliability and power security. Prior to uprating, need of transmission of more amount of power is felt. Analysis is made to see whether need can be fulfilled by uprating of overhead lines without hampering stability and security.

10.6.6 Upgrading

Upgrading of overhead lines refers to an enhancement of capability limits of the line with respect to thermal, mechanical and electrical boundaries considering all constrains involved in particular line. Upgrading may involve with an upgradation of insulation levels and conductor. In few cases, it may involve with a change of support or support structure. Cost involved in upgrading is higher than uprating.

10.6.7 Conversion

Conversion of existing line into new category is recommended to reduce loss (both electrical and financial), risk, power theft, failure, etc. and to improve security, stability, utility, capability, etc. For example, overhead line with a bare conductor for low-voltage distribution is often converted into overhead line with a covered conductor or into underground line. In transmission system, lines of particular voltage level are converted into lines of a greater voltage level or EHV AC lines are converted into HVDC lines. The conversion of old line has the advantage of using the same path except some exception where the alternate path may have to be chosen for a particular span for area-specific reasons.

10.6.8 Extension

Extension of overhead lines is required mainly for the following reasons:

- Increase of power demand
- Increase of demand of old consumer
- Insertion of new consumer
- Increase of coverage area of distribution
- In parallel with conversion of old line

Overhead lines face extension with respect to length and capacity. Capacity may be increased by adding more circuits or by parallel line in the same path of old overhead lines. Extension goes through the following steps: fixing aim, proposal, approval, design and approval, manufacturing, construction, tests and commissioning. For extension, approval and agreement with owner(s) must be done before execution. One advantage of extension over setting up new line is that establishment of a fully new line needs path selection whereas the extension of line reduces the job of path selection to a great extent. Only the extension part is brought under main consideration in path selection reducing the overall cost.

10.7 Research direction

A lot of research works are happening on route selection, planning and construction, commission, operation and maintenance. Different reviews and experiences are being shared. Standards are being updated and new guidelines by International Electro-Technical Commission, IEEE, etc. are being introduced. For the completion of project work, different project management solutions have been introduced. Different advanced and effective tools of optimization have been applied in planning and execution of project of making overhead lines.

Different types of modern management system are being introduced. Advanced technical tools are used for monitoring. For example, radar-based satellite monitoring is being used for monitoring areas of 'right-of-way' in risk management. Use of satellite monitoring has made monitoring possible and easier than the previous ones for the areas which are located at remote end and at geographically critical areas. For risk management, there is no specific mathematical formula. However, different fuzzy, artificial neural network based tools are used for the prediction of risk in overhead lines [7].

A vehicle-based mobile manipulator has been used in a 6.6 kV distribution line [8]. The system has been found effective to work for construction and maintenance in live line. It has three main sections: truck, insulated boom and working station. Based on the type of operator, two types of mobile manipulators are found. In the first type, operator acts in live line sitting at the working station located above insulated boom. He takes all safety measures and insulation to work in live condition and works by controlling manipulator which has normally 6 degrees of freedom for movement. In the second type, operator works from ground level. Three-dimensional camera, robots and manipulators are installed in working station having 6 degrees of freedom for movement. Operator monitors through a camera and operates with the help of an automated robot from ground level. Robotics applications have been found effective in live line maintenance work in power distribution as well as in transmission line [9].

10.8 Summary

This chapter has presented various aspects of route selection, planning and construction, commission, operation and maintenance. Different steps involved in setting up overhead line before going into operation have been presented. It shows that after setting up objective, route selection will finalize the path of overhead lines. The planning is done. It is followed by construction work. It involves foundation, installation and erection. Protective and safety measure must be incorporated in overhead lines. Smooth operation is expected, and regular maintenance will ensure long-life span of the overhead line.

References

[1] K. O. Papailiou, Overhead Lines. CIGRE Green Books, Springer, Malters, Switzerland; 2017.

[2] Overhead Conductor Installation Guide Recommended Practices. 1st Edition, Installation guide by Electric Utility Engineering Section. General Cable Technologies Corporation, Kentucky, USA; 2014.

[3] P. N. Micropoulos and T. E. Tsovilis, 'Lightning attachment models and perfect shielding angle of transmission lines'. 44th Universities Power Engineering Conference, Glasgow, Scotland, Paper Number 57-70151, 2009.

[4] IEEE Std. 1243-1997: IEEE Guide for Improving the Lightning Performance of Transmission Lines; December 1997.

[5] R. Hileman, "Shielding of Transmission Lines," Insulation Coordination of Power Systems. CRC Press, Taylor & Francis Group, New York, NY, USA; 1999, pp. 1–6.

[6] R. Minguez, R. Martinez, M. Manana, A. Arroyo, R. Domingo, and A. Laso, 'Dynamic management in overhead lines: a successful case of reducing restrictions in renewable energy sources integration'. in Electric Power System Research, vol. 173, pp. 135–142, 2019, ISSN 0378-7796, doi:10.1016/j.epsr.2019.03.023.

[7] K. Hafeez, 'Risk management of overhead electric power lines'. in International Journal of Power System, vol. 3, pp. 53–57, 2018.

[8] H. Yokoyama, M. Mukaida, A. Uchlyama, Y. Tamura and Y. lwaki, Manipulator System for Constructing Overhead Distribution Lines, 0-7803-1340-2/93, IEEE, MB6327, 1993, pp. 445–457.

[9] J. H. Dunlap, J. M. Van Name, and J. A. Henkener, 'Robotic maintenance of overhead transmission lines'. in IEEE Transaction on Power Delivery, vol. PWRD-1, no. 3, pp. 280–284, 1986.

Further reading

[1] P. Pelacchi, 'Automatic hot line insulator washing device positioned by helicopter'. ESMO'98 – 1998 IEEE 8th International Conference on Transmission and Distribution Construction, Operation and Live-Line Maintenance Proceedings ESMO'98 Proceedings, ESMO 98 The Power is in Your Hand, Orlando, FL, USA, 1998, p. 133, doi:10.1109/TDCLLM.1998.668343.

[2] M. Boschetti, 'Conversion with live maintenance of the medium tension power lines (15 kV) from rigid insulators into string insulators'. 2003 IEEE 10th International Conference on Transmission and Distribution Construction, Operation and Live-Line Maintenance, 2003, 2003 IEEE ESMO, Orlando, FL, USA, 2003, pp. 181–185, doi:10.1109/TDCLLM.2003.1196484.

[3] B. Avidar, 'Computerized design of overhead transmission power lines'. Proceedings of ESMO'93, IEEE 6th International Conference on Transmission and Distribution Construction and Live-Line Maintenance, Las Vegas, NV, USA, 1993, pp. 199–203, doi:10.1109/TDCLLM.1993.316247.

[4] M. Alasmry, K. Alsayed, J. Alzahrni, and A. Hobany, 'Reducing the duration of right-of-way acquisition process for high voltage transmission power lines projects'. 2016 Saudi Arabia Smart Grid (SASG), Jeddah, 2016, pp. 1–7, doi:10.1109/SASG.2016.7849672.

[5] E. E. F. Creighton, 'Questions on the economic value of the overhead grounded wire'. in Journal of the American Institute of Electrical Engineers, vol. 41, no. 1, pp. 21–29, 1922, doi:10.1109/JoAIEE.1922.6594378.

[6] S. P. Simon, N. P. Padhy, J.-B. Park, *et al.*, 'Power system planning and operation'. Applications of Modern Heuristic Optimization Methods in Power and Energy Systems. Wiley-IEEE Press, Chichester, West Sussex, UK; 2020, pp. 39–225, doi:10.1002/9781119602286.ch3.

[7] M. Eremia, C.-C. Liu, and A.-A. Edris, 'CSC–HVDC transmission'. Advanced Solutions in Power Systems: HVDC, FACTS, and Artificial

Intelligence. Wiley-IEEE Press, Chichester, West Sussex, UK; 2016, pp. 35–124, doi:10.1002/9781119175391.ch3.

[8] Q. Huang, A. H. Khawaja, Y. Chen, and J. Li, 'Magnetic field measurement for power transmission systems'. Magnetic Field Measurement with Applications to Modern Power Grids. Wiley-IEEE Press, Chichester, West Sussex, UK; 2019, pp. 53–142, doi:10.1002/9781119494485.ch3.

[9] M. A. Ibrahim, 'Case studies related to overhead transmission-line system disturbances'. Disturbance Analysis for Power Systems. Wiley-IEEE Press, Chichester, West Sussex, UK; 2012, pp. 461–570, doi:10.1002/9781118172094.ch6.

[10] ANSI/IEEE Std 516-1987: IEEE Guide for Maintenance Methods on Energized Power-Lines; 1988, pp. 0_1, doi:10.1109/IEEESTD.1988.81528.

[11] IEEE Std 524-2016 (Revision of IEEE Std 524-2003): IEEE Guide for the Installation of Overhead Transmission Line Conductors; 28 April 2017, pp. 1–162, doi:10.1109/IEEESTD.2017.7912224.

[12] IEEE Std 664-1980: IEEE Guide on the Measurement for the Performance of Aeolian Vibration Dampers for Single Conductors; 16 October 1980, pp. 1–14, doi:10.1109/IEEESTD.1980.81009.

[13] IEEE Std 738-1993: IEEE Standard for Calculating the Current-Temperature of Bare Overhead Conductors; 8 November 1993, pp. 1–48, doi:10.1109/IEEESTD.1993.120365.

[14] IEEE 751-1991: IEEE Trial-Use Design Guide for Wood Transmission Structures; 20 February 1991, pp. 1–119, doi:10.1109/IEEESTD.1991.7407693.

[15] IEEE Std 977-2010 (Revision of IEEE Std 977-1991): IEEE Guide to Installation of Foundations for Transmission Line Structures; 15 February 2011, pp. 1–87, doi:10.1109/IEEESTD.2011.5716533.

[16] ANSI/IEEE Std 987-1985: IEEE Guide for Application of Composite Insulators; 1985, pp. 0_1, doi:10.1109/IEEESTD.1985.82458.

[17] IEEE Std 998-1996: IEEE Guide for Direct Lightning Stroke Shielding of Substations; 15 February 1996, pp. 1–176, doi:10.1109/IEEESTD.1996.81546.

[18] IEEE Std 1025-1993: IEEE Guide to the Assembly and Erection of Concrete Pole Structures; 28 December 1993, pp. 1–32, doi:10.1109/IEEESTD.1993.8241473.

[19] IEEE Std 1410-1997: IEEE Guide for Improving the Lightning Performance of Electric Power Overhead Distribution Lines; 31 December 1997, pp. 1–48, doi:10.1109/IEEESTD.1997.85324.

[20] IEEE Std 1628-2009: IEEE Recommended Practice for Maintenance of DC Overhead Contact Systems for Transit Systems; 25 September 2009, pp. 1–58, doi:10.1109/IEEESTD.2009.5271980.

[21] IEEE Std 1724-2011: IEEE Guide for the Preparation of a Transmission Line Design Criteria Document; 27 May 2011, pp. 1–32, doi:10.1109/IEEESTD.2011.5783877.

[22] IEEE Std 1829-2017: IEEE Guide for Conducting Corona Tests on Hardware for Overhead Transmission Lines and Substations; 31 March 2017, pp. 1–32, doi:10.1109/IEEESTD.2017.7891097.

[23] IEEE Std 1863-2019: IEEE Guide for Overhead AC Transmission Line Design; 11 May 2020, pp. 1–109, doi:10.1109/IEEESTD.2020.9086170.
[24] IEEE Std 1893-2015: IEEE Guide for the Measurement of DC Transmission Line and Earth Electrode Line Parameters; 25 February 2016, pp. 1–37, doi:10.1109/IEEESTD.2016.7419218.
[25] IEEE Std C135.63-1998(R2006): IEEE Standard for Shoulder Live Line Extension Links for Overhead Line Construction; 5 December 2006, p. i, doi:10.1109/IEEESTD.1998.88264.
[26] ANSI C135.31-1988: American National Standard for Zinc-Coated Ferrous Single and Double Upset Spool Insulator Bolts for Overhead Line Construction; 12 September 1988, pp. 1–4, doi:10.1109/IEEESTD.1988.600606.
[27] ANSI C135.22-1988: American National Standard for Zinc-Coated Ferrous Pole-Top Insulator Pins With Lead Threads for Overhead Line Construction; 3 January 1992, pp. 1–4.
[28] IEEE Std C2-2007: National Electrical Safety Code 2007 Edition; 1 August 2006, pp. 1–336, doi:10.1109/IEEESTD.2006.322219.
[29] D. J. Marne, McGraw-Hill's National Electrical Safety Code® (NESC®) 2017 Handbook. McGraw-Hill Education; 2017.

Index

www.ingramcontent.com/pod-product-compliance
Lightning Source LLC
LaVergne TN
LVHW011842200926
848484LV00002B/387